T0140142

China Internet Development Report 2018

Chinese Academy of Cyberspace Studies

China Internet Development Report 2018

Blue Book of World Internet Conference

 Springer

Chinese Academy of Cyberspace Studies
Beijing, China

Translated by
Peng Ping
Beijing Foreign Studies University
Beijing, China

ISBN 978-981-15-4045-5   ISBN 978-981-15-4043-1   (eBook)
https://doi.org/10.1007/978-981-15-4043-1

Jointly published with Publishing House of Electronics Industry
The print edition is not for sale in China (Mainland). Customers from China (Mainland) please order the
print book from: Publishing House of Electronics Industry.

Foreword

Today, information technology is changing day by day, represented by digitalization, networking, and intelligence, contributing to the development of economy and society, national governance systems and governance capacity, and the people's increasing demand for a better life. Thanks to this historic opportunity, China's Internet has entered the fast lane of development. Especially, the country has worked out a road of Internet development and governance with Chinese characteristics and has contributed to the world Internet development.

Against this backdrop, we have compiled *China Internet Development Report 2018* (referred to as "the *Report*" hereinafter), in which the compiler tries to study and analyze in detail China's Internet development by summarizing what the country has done, analyzing the status quo, and looking into the future, so that the *Report* will be the best choice for people to know about the past, present, and future of China's Internet. To this end, the compiler tries to:

1. We take General Secretary Xi Jinping's Strategic Thoughts on Cyber Power as the Theoretical Basis and Guideline for the Compilation

Since the 18th National Congress of the Communist Party of China (CPC), General Secretary Xi Jinping, a visionary man, has put forward a series of new ideas, thoughts, and strategies based on China's Internet development and governance, systematically illustrating the important theoretical and practical issues concerning cybersecurity and informatization and thus forming rich, profound, scientific, and systematic strategic thoughts on cyber power. All these ideas, thoughts, and strategies are an important part of Xi's Thought on Socialism with Chinese Characteristics for a New Era. In particular, in the speech he made at the National Cybersecurity and Information Technology Conference, based on the general situation of the CPC and the country, he scientifically analyzed the trend and mission of information reform, systematically illustrated his thoughts on cyber power, and answered a series of theoretical and practical questions about cyber affairs. It is the guideline for making China into a cyber power. We take Xi Jinping's strategic thoughts on cyber power as the spirit and main thread running through the *Report* to interpret Xi's ideas, philosophy, and thoughts so that the readers can accurately grasp the significance of his thoughts.

2. We Take the Practice of China's Internet Development as the Research and Reality Basis

Guided by Xi Jinping's Thought on Socialism with Chinese Characteristics for a New Era, especially his strategic thoughts on cyber power, the country has achieved a lot in Internet development and governance. The cyberspace is becoming clearer day by day, the national cybersecurity defense is being consolidated, informatization's role in boosting the economic and social development is becoming prominent, China's power of discourse and influence in cyberspace keeps increasing and the Chinese people are having more sense of gain in sharing the benefits of Internet development. All these vivid achievements provide broad space and rich resources for the compilation of the *Report*. While illustrating the status quo of China's Internet development, and new practice and achievements in particular, the compiler tries to show China's experience in Internet governance.

3. We Take Comprehensiveness, Accuracy, and Objectivity as the Aim and Principle for the Compilation

In the process of compilation, we always follow the principle of making everything internationalized, authoritative, accurate, theoretical, and generalized, to show the Chinese academic circle's understanding of and thought on the Internet development. We try to make the Report comprehensive by doing an all-field and panoramic research on China's network infrastructure, information technology, digital economy, e-governance, network media, cybersecurity and Internet law construction, and international cyberspace governance; we try to make all the data accurate, authoritative, and up-to-date by collecting data from governmental agencies, industries, and research institutes; we try to make our assessment objective by setting up China's Internet Development Index on the basis of the mature index systems from both home and abroad, making assessment on cybersecurity and informatization in 31 provinces (autonomous region and municipalities directly under the Central Government) from 6 dimensions to comprehensively and accurately reflect the Internet development level throughout the country.

It is our sincere hope that the *Report* will provide a new drive for China's Internet development, and will be used as a new window for the world to know about China's Internet development, and a new reference for the world Internet development.

Beijing, China   China Academy of Cyberspace
October 2018

Contents

Overview

Today, information technology (IT) represented by the Internet is developing day by day. Digitalization, network transformation, and intelligent development are playing an increasingly important role in boosting economic and social development and modernization of the national governance system and capability, and in meeting the people's ever-growing needs for a better life. Facing the historic opportunities brought about by IT development, the Communist Party of China (CPC) and the Chinese Government both attach importance to Internet development, application, and governance, coordinating major issues of cybersecurity and informatization in politics, economy, culture, society, and military affairs. They have made important decisions and launched important measures to promote cyberspace administration, which has witnessed historic achievements.

The year 2018 is the opening year of comprehensively carrying out the spirit of the 19th National Congress of CPC, the 40th anniversary of the country's adoption of the reform and opening-up policy, and the decisive and transitional year of building a moderately prosperous society in all respects and of implementing the 13th Five-Year Plan. It is also an important year for China's cyberspace administration to enter a new era, in which it will meet new challenges and see new progress. In March 2018, the Central Leading Group for Cyber Affairs was renamed the Central Committee for Cyber Affairs in accordance with *Decision on Deepening Reform of Party and State Institutions*. It is in charge of cyber affairs, including top design, overall layout, coordination, supervision, and implementation of policies. Thus, the responsibilities of the Office of Cyber Security has been optimized. From April 20th to 21st, the National Cyber Affairs Conference was held in Beijing. General Secretary Xi Jinping delivered an important speech, in which, he, by starting from the overall situation of the CPC and the country, analyzed the trend and mission of the IT revolution, systematically illustrated the important thoughts on cyber power and answered a series of major theoretical and practical questions. His speech provides guidance and fundamental principles for building China's strength in cyberspace. Guided by Xi Jinping Thought on Socialism with Chinese Characteristics for a New Era and especially his thoughts on cyber power, the Internet of China, in accordance with the strategic layout made at the 19th National

Congress of the Communist Party of China, is seeing rapid development. Great achievements have been made in cyber strength, digital development, and smart society, contributing to and providing a guarantee for the development of the CPC and the country.

I. General Secretary Xi Jinping's Thoughts on Cyber Power Provide Guidance and Blueprint for China's Cyber Administration Development

Since the 18th National Congress of the Communist Party of China (CPC), General Secretary Xi Jinping, a visionary man, has put forward a series of new ideas, thoughts, and strategies based on China's practice of Internet governance. All of them are systematic illustrations on important theoretical and practical issues concerning cybersecurity and informatization, covering cyber content building, cyber security, informatization, and international cyber governance, and guiding China's cyber administration in witnessing historic achievements and revolution. An Internet governance path with Chinese characteristics and Xi Jinping's thoughts on cyber power have been formed.

The important speech General Secretary Xi Jinping delivered at the National Cyber Affairs Conference gives systematic illustrations of his thoughts on cyber power. He made explicit the important position of cyber administration in the general layout of the CPC and national development and the strategic goal and tasks of and principles for building China's strength in cyberspace, and put forward proposals on international Internet governance, along with basic methods for cyber administration. The thoughts conform to the trend of the world Internet development, revealing historic opportunities and challenges brought about by the information revolution to China's economic and social development. A series of fundamental and strategic questions are answered concerning the direction and overall situation of the country's cyber development. Therefore, they are the guidelines for building China's strength in cyberspace.

According to Xi Jinping, we should construct a concentric circle of online and offline communities and reach a better consensus to consolidate the ideological foundation for forging ahead in unity. Cyberspace is the common home of humankind and the new space for people's learning, working, and life and a new platform for obtaining public services. Making it better, cleaner, and safer accords with the interests of the people. To enhance Internet governance and content building requires the improvement of comprehensive cyber governance capability, and the CPC committees' leadership, governmental administration, businesses' accountability, social supervision, and Internet users' self-disciplining, as well as the combination of economic, legal, and technical means. We must enhance positive publicity, adhering to the right directions of politics, public opinions, and values, and uniting all the Internet users with Xi Jinping Thought on Socialism with

Chinese Characteristics for a New Era and the spirit of the 19th National Congress of the Communist Party of China. We should carry out education on ideals and convictions, deepen the publicity of Socialism with Chinese Characteristics for a New Era and the China Dream and cultivate and practise the core socialist values. We should administrate, open, and visit the websites in accordance with laws, enhance the accountability of Internet businesses and self-disciplining of the Internet industry, give full play to Internet users' initiative, and mobilize all sectors to participate in Internet governance.

Xi points out that without cybersecurity there would be no national security, stable economy, and society, or people's guaranteed interests. There are an increasing number of threats to and risks in cyber security, which are penetrating into politics, economy, culture, society, ecosystem, and national defense. Anything slight in cybersecurity may affect the whole situation, since it is a significant issue concerning national security and development as well as the people's interest. To safeguard cybersecurity requires the right outlook on it and the realization of the fact that it is something as a whole rather than separate, something dynamic rather than static, something open rather than closed, something relative rather than absolute, and something shared rather than isolated. We must improve our capacity in cybersecurity protection as soon as possible, and strengthen information infrastructure security protection, and construction of the mechanism, means, and platforms for cybersecurity information coordination. We must build the capacity in guiding cybersecurity emergency response, develop cybersecurity industries, and prevent any hazards in advance. Insisting that the cybersecurity should serve and rely on the people, we will implement the accountability mechanism, combat cyber-attack, online fraud, and infringement on citizens' privacy, and carry out popularization of cybersecurity knowledge and skills.

According to Xi Jinping, core technologies are the pillar of a nation. Therefore, we must make up our mind, keep our perseverance, and find the focus to speed up breakthroughs in core technologies of information. Network information technology is an area of technical innovation with the highest input in R&D, the most active innovation, the most extensive application, and the most remarkable radiating significance. It is the height of competition in global technical innovation. So we must realize that China has to catch up with leading countries in core technology breakthroughs. As the "life-gate" for a country's development, core Internet technologies being controlled by others will be a hazard for us. To speed up breakthroughs in them is the fundamental solution to cybersecurity and the requirement of cyber administration development. We should construct the industrial system involving technology, industry, and policy-making while improving the institution and market favorable to innovation and entrepreneurship and reinforcing intellectual property protection and vitalizing innovation. We should provide green channels for the combination of basic research and technical innovation, trying to promote team breakthroughs of applied technologies through basic research.

According to General Secretary Xi Jinping, the cyber and information technology sector represents the new productivity and new development direction, so it is expected to take one step ahead in practising the new development philosophy.

With the accelerated penetration of technical innovation into all areas of economy and society, informatization is producing fundamental and comprehensive influence on the operation of economy and society as well as on production and lifestyles, becoming a driving force for economic transformation and upgrading and high-quality development. To practise the new philosophy of development requires the cultivation of new driving forces through informatization and promotion of new development through new driving forces. We should strengthen network infrastructure construction and in-depth information resource integration, and open up the "main artery" of information for economic and social development. We should develop digital economy, accelerate digital industrialization, drive innovation through IT, and produce new business forms and new modes. We should promote industrial digitalization, transform traditional industries from all dimensions, all angles, and all chains, and speed up digitalization, networking, and intelligent development of manufacturing, agriculture, and service industries, so that we can improve the total factor productivity and data can play an amplifying, superimposing, and multiplying role in the economic development.

Xi points out that to promote the global Internet governance is the general trend conforming to the people's will. In the information era, the Internet is making the international community into a community of shared future and giving rise to new challenges to national sovereignty, security, and development interests. How to govern the Internet and make good use of it is a common problem faced with the international community. Xi has proposed "four principles" for promoting the global Internet governance system reform and "five proposals" for constructing a community of shared future in cyberspace, both having won more and more international recognitions. Internet cyber governance should involve all sides and all parties, including governments, international organizations, Internet businesses, technical communities, civil organizations, and individual citizens. We should take the security of all countries into consideration, expand cooperation, and innovate the ways of cooperation to jointly promote development, protect security, and participate in governance and share achievements. In particular, we should take the construction of the Belt and Road as a good opportunity to strengthen developing countries' cooperation in network infrastructure construction, digital economy, and cyber security, so that we can build the Digital Silk Road of the 21st century.

General Secretary Xi Jinping points out that the Internet development must be oriented to benefit the people, started from and aimed for the people's well-being, so that they will have more sense of gain, happiness, and safety in enjoying the benefits of Internet development. Since its access to the Internet, China has been taking the benefit of its population of 1.3 billion as the most important and trying to enable all the people to enjoy the Internet development achievements. Thanks to the people's participation and support, the Internet of the country has witnessed permanent healthy development and prosperity; owing to their increasing expectation and demand, the Internet of China has got its direction and goal of development. To make the Internet better serve the people, we should speed up the popularization of informatized services and reduce the use cost, so that all can afford to use the Internet and then have the chance to use it and use it well. We should adopt the mass line in developing the Internet, trying to have government and CPC affairs open through informatization, speed up e-government building, and construct full-process, one-stop online service platforms. We will promote "Internet + education", "Internet + medical care", and "Internet + culture" to make services accessible, beneficial, and convenient to all people, who thus can enjoy services with little effort through the Internet.

General Secretary Xi Jinping's thoughts on cyber power are the scientific results of the CPC's theoretical and practical innovation, and creative answers to a series of major questions concerning informatization through the full use of Marxist position and viewpoints. They form an important part of Xi Jinping Thought on Socialism with Chinese Characteristics for a New Era. They are fundamental guidelines for cyberspace administration, of profound significance to building China's strength in cyberspace and constructing digital China and smart society. We must keep carrying them out and enriching them.

II. Achievements and Advances of China's Internet Development in 2018

In 2018, guided by Xi Jinping Thought on Socialism with Chinese Characteristics for a New Era, and especially his thoughts on cyber power, the country, by seizing the historic opportunity of information development, has enhanced content building, safeguarded cyber security, and sped up IT R&D while promoting the economic and social development through informatization and participating in international Internet governance. New achievements and advances have been made.

1. China's information infrastructure keeps being upgraded and has become the strategic foundation for economic and social transformation.

Thanks to the Broadband China Strategy, the Action of Facilitating Faster and More Affordable Internet Connection and the Pilot Project of General Telecommunication Service, China's information network infrastructure has witnessed leapfrog development. Specifically, fiber broadband, 4G networks, cloud computing, and Content Delivery Networks (CDNs) keep being upgraded, 5G technology is witnessing accelerated R&D and industrial layout, IPv6 is to be deployed and commercially used in a large scale and network service quality has been improved, which lays a solid foundation for economic and social transformation. By June 2018, there were 378 million fixed broadband users, including 328 fiber broadband users, accounting for 87.5% of the total number of broadband users and ranking first in the world. Demonstration of Gbps broadband has been done in full swing, and China is welcoming its fiber era nationwide. The web download speed has been increased dramatically. By the end of the second quarter of 2018, the download speed of fixed broadband and 4G networks of the country had surpassed 20 Mbps, both with a year-on-year increase of over 50%. The country has caught up in the construction of

4G networks and even surpassed some countries in that aspect. There are over 3.4 million 4G base-stations, and 1.11 billion 4G users throughout the country, with the 4G penetration rate being among the world's top five. China is one of the leading countries in the R&D of 5G technology. It is constructing the world's biggest 5G testing ground. The broadband networking capacity and coverage have been improved in rural areas, with the coverage rate in administrative villages amounting to 97.4%, that of fiber broadband to 96%, and that of 4G networks to 95%. The quantity of critical Internet resources has seen steady growth, with the number of IPv4 addresses amounting to 338.82 million, and that of IPv6 to 23,555 (in terms of blocks)/32 (in terms of addresses), both ranking second in the world. The international exit broadband traffic is 8,826.302 Gbps, with a year-on-year growth of 10.68%.

2. China has seen progress in Internet information technology and made single-point breakthroughs in some areas.

Network information technology is witnessing its rapid development. By seizing the opportunity brought about by the new-round technical revolution and relying on self-dependent innovation of core technologies, China is speeding up the layout of cutting-edge and nonsymmetrical technologies, and through technical, industrial, and policy support, promoting progress in the R&D and application of high-performance computing, mobile communication, quantum communication, core chips, and operating system. In 2017, the country applied for 3.986 million patents in IT, including 48,900 PCT patents, ranking second in the world for the first time.[1] According to Global Innovation Index 2018 released by WIPO and other agencies, China's innovation capacity keeps improving and its innovation index ranks 17th, listed into the top 20 for the first time. By June 2018, it had created a national high-performance computing service environment made up of 17 high-performance computing centers, with its resource capability among the top of the world. In July 2018, Tianhe No. 3 E-class prototype was successfully developed, with its computing capability 200 times higher and its storage capacity 100 times higher than that of Tianhe No. 1. Internet businesses are key forces in IT R&D, accelerating the application of AI, cloud computing, and big data and the emergence of a large number of innovative intelligent service scenarios. For example, Baidu's PaddlePaddle (a deep learning platform), Apollo's automatic driving platform, Alibaba's ET City Brain, and Tencent's auxiliary medical diagnosis and treatment platform are all influential AI software platforms. They have been put into application in many scenarios, such as smart city, medical care and health, education and amusement, and housekeeping and elderly care. On the other hand, China still lags behind countries leading in IT, with weaknesses and shortcomings in some areas, so we have to accelerate deployment and promote development in such areas.

---

[1]Source: *Information Technology Patents* (2018), China Industrial Control Systems Cyber Emergency Response Team (CICS-CERT).

3. China's digital economy is gathering momentum, contributing to the formation of the new development mode in which data and real economy are deeply integrated.

China adheres to cultivating the new drive through informatization and promoting new development through the new drive. New economy represented by digital economy is thriving, boosting economic and social development. In 2017, the digital economy scale of the country reached RMB 27.2 trillion *yuan*, accounting for 32.9% of the national GDP, with a year-on-year growth of 2.6%, and digital economy contributed 55% to the GDP. In the same year, e-commerce transaction volume amounted to RMB 29.16 trillion *yuan*, with a year-on-year growth of 11.7%; social e-commerce witnessed rapid development, with the number of social retail users amounting to 223 million, contributing to the expansion of e-commerce. There were 171 million employees in digital economy, accounting for 22.1% of the total number of employees that year. Therefore, digital economy is an important channel for employment. Digital industrialization has witnessed rapid development, technical application is commercialized at a high speed and new industries and new business forms keep emerging. In 2017, the scale of IoT market was over RMB one trillion *yuan*, with a compound annual growth rate of over 25%; the revenue of key IoT businesses that have gone public was RMB 483.38 billion *yuan*, with a year-on-year growth of 20.7%[2]; the AI market volume was RMB 23.7 billion *yuan*, with a year-on-year growth of 67%.[3] Industrial digitalization keeps being deepened and the penetration rate of digital economy in industry, agriculture, and service is, respectively, 17.2, 6.5, and 32.6%. Intelligent manufacturing projects are being implemented. By the end of the first quarter of 2018, the popularization rate of digital R&D design tools in industrial businesses above the designated size was 67.4%, and the numerical control rate of critical process was 47.8%. Coordinative networked manufacturing, personalized customization, and service-oriented manufacturing are new highlights in traditional industries. China keeps optimizing the market environment and policies for Internet businesses, and a number of such businesses with innovation vitality and development potential have been cultivated, with their international competitiveness increased. By December 2017, there were 102 Chinese Internet businesses that had gone public at home or abroad, with a growth of 12% in comparison to that in 2016. There are 77 unicorn Internet businesses,[4] and such businesses with the scale of over RMB one hundred million *yuan* keep emerging. They are brilliant stars promoting the development of digital economy.

---

[2]Source: *Annual Report on China's IoT Development 2017–2018*, China Economic Information Service.

[3]Source: *AI Development Report2018*, Tsinghua University.

[4]Source: *The 41st China Statistical Report on Internet Development*, CNNIC.

4. China's cyber protection capacity keeps improving dramatically, and a solid cybersecurity shield has been set up.

China is accelerating the construction of its cybersecurity system, having improved its cybersecurity guarantee capacity and level and formed an overall cybersecurity defense line to tackle relevant threats in the new situation. Over the past year, the country has reinforced the protection of critical information infrastructure, carried out cybersecurity examinations of all critical information infrastructure, and checked the risks of cybersecurity concerning such infrastructure of key areas and industries to enhance the capability of cybersecurity protection. Technologies of cybersecurity keep being improved. A large quantity of such technologies based on big data, AI, and blockchain are becoming mature, witnessing extensive use in cybersecurity protection. Regulations supporting *Cybersecurity Law* are being formulated and cybersecurity censorship has been established. China is speeding up the standardization of cyber security, having issued *Information Technology—Personal Information Security Specification* and other important national standards. In September 2018, the Fifth National Cybersecurity Publicity Week was successfully held. Through a host of activities like cybersecurity Exposition and Internet Literacy, it has contributed to the improvement of the cybersecurity protection awareness and skills of the public. The scale of cybersecurity industry has increased. In 2017, it was RMB 43.92 billion *yuan*, with a year-on-year growth of 27.6%; the number of cybersecurity businesses was over 2,681.[5] The international competitiveness of China's cybersecurity products and services has been enhanced.

5. China's cyberspace is becoming cleaner day by day and its cyber culture enjoys prosperity.

The cyberspace is the common home of China's Internet users. Positive energy publicity is the general requirement, and controllability is the absolute principle. The country is enhancing content building, launching projects such as Good Internet Users in China and Network Media Reporting the Grass Roots, as well as a large number of phenomenal news broadcasts and thematic publicity products, and spreading and introducing core socialist values. The capability of excellent cyber content supply keeps being enhanced, and communities of digital content consumers have been cultivated. By June 2018, the number of network video users was 609 million; that of network music listeners, 555 million and that of network literature readers, 406 million. Besides, the environment for digital content copyrights has been improved.[6] Legislation on cyber governance is being sped up. In 2018, China launched *the E-commerce Law of the People's Republic of China* as the legal basis for Internet governance. People take an active part in comprehensive Internet governance, and http://www.piyao.org.cn/, a united rumor refuting platform, has been put into use, adopting the work pattern of linked discovery, tackling,

---

[5]Source: *White Paper of China's cybersecurity Industry Development 2018*, China Academy of Information and Communications Technology.
[6]Source: *The 42nd China Statistical Report on Internet Development*, CNNIC.

and rumor refuting. In the first half of 2018, 39.028 million complaints concerning cyber space were tackled, with a year-on-year growth of 117.1% in complaint handling. Harmful information can be detected and tackled timely, so the cyberspace is becoming cleaner day by day.

6. China's international communication and cooperation in cyberspace are being deepened, and it is continuously contributing Chinese experience and wisdom to the world Internet development.

China takes an active part in Internet governance, strengthens international communication and cooperation, and works for joint promotion, protection, participation, and result sharing. The country has successfully hosted five sessions of the World Internet Conference, which is an international platform for cooperation, leading to a series of influential cooperative achievements made by Chinese and other governments, social organizations, and businesses, including the signing of a number of cooperative agreements such as the *Belt and Road Initiative for International Digital Cooperation*. China is deepening its cooperation with the United States, Russia, and Europe, and expanding its cooperation with emerging markets and developing countries in cyberspace, to combat cybercrime, boost digital economy, and accelerate infrastructure construction. In September 2018, the Third China-ASEAN Information Harbor Forum was held. Informatization is the focus of the cooperation between China and ASEAN countries. Besides, cooperation has been promoted among countries along the Belt & Road, and China offers advanced experience and practical solutions in Internet development to more countries.

7. China's Internet development better meets the expectations and needs of the people, whose sense of gain has dramatically increased.

As has been mentioned, China's development is always oriented to benefit the people, started from, and aimed for their well-being, so that they will have more sense of gain, happiness, and safety in enjoying the benefits of Internet development. By June 2018, the number of China's Internet users had reached 802 million and that of websites, 544 million, and the Internet popularization rate, 57.7%. Information service is becoming faster and more convenient, in better quality and lower price. The average expenditure on fixed band and mobile traffic has decreased by 90% in comparison with that in 2014, and that on dedicated Internet access has decreased by 30%. Long-distance charge and roaming charge for mobile calls have been abolished. Traffic of mobile Internet access has seen explosive growth. It witnessed an accumulated growth of 26.6 billion GB in the first 6 months of 2018, with a year-on-year growth of 199.6%.[7] "Mobile life" is people's first choice. Poverty alleviation through Internet serves as a new way of targeted poverty alleviation and elimination. E-commerce is penetrating into impoverished counties through its comprehensive demonstration in rural areas. 499 national-level

---

[7]Source: *The 42nd China Statistical Report on Internet Development*, CNNIC.

impoverished counties have been channeled into the support, accounting for 60% of all impoverished counties. Thus, the income of 2.74 million people from registered impoverished households has increased. Informatization has been adopted to improve the government's service efficiency. The number of online governmental service users had reached 470 million by June 2018. The rate of using online governmental service was 42.1% through Alipay and the WeChat City Service Platform, which are the most frequently used online government service platforms. People can enjoy services with little effort through the Internet, which is a vivid embodiment of China's governmental service in the new era.

III. Remarkable Achievements of Innovative Internet Development in all Provinces, Autonomous Regions, and Municipalities Directly Under the Central Government

In *China Internet Development Report 2017* released last year, China's Internet Development Index was set up and published for the first time, covering infrastructure construction, innovation capacity, development of digital economy, Internet application, cyber security, and cyberspace administration, which were used to assess and rank the effect and level of Internet development in all provinces, autonomous regions, and municipalities directly under the Central Government ("municipalities" for short hereinafter). It is a reference for all regions to seize the development opportunities and catch up in some fields. This year, we have improved the Internet Development Index and followed up the latest effects and progress in Internet development of the province, municipalities, and autonomous regions, to assess and reflect their Internet development more comprehensively, objectively, and scientifically.

1. China's Internet Development Index has been optimized and improved.

To make a better assessment on the Internet development throughout the country, we have optimized and improved China's Internet Development Index (see Table 1). We have adjusted and improved the index system, optimized the algorithm model, and enhanced data collection, so that the assessment is more authoritative, scientific, and accurate.

Table 1 China's Internet Development Index

Indicator	Key assessment factors	Weight	Specification
Infrastructure construction	Number of broadband access ports and percentage of optical fiber users, number of 4G mobile base stations and percentage of 4G users, and number of IDC centers	10%	Local infrastructure construction level and Internet coverage of broadband, mobile and wireless networks, and cloud computing
Innovation capacity	Number of patents registered, index of human resources in information society, and investment in R&D	20%	Level, capacity, and environment of local industrial innovation and local talents cultivation
Development of digital economy	ICT industry, income of telecommunication, e-commerce, and development of Internet businesses	20%	Development of local ICT industry, e-commerce, and development of Internet businesses
Internet application	Rate of Internet coverage, scale of e-commerce, online full-process government service rate, and online full-process public service rate	25%	Local individuals' use and corporations' use of the Internet, e-government, and public service application
Cyber security	Malicious computer program, number of controls against web vulnerabilities, number of cybersecurity businesses, and talents of cyber security	13%	Local security of Internet and websites, as well as cybersecurity industry and talent production
Cyber administration	Number of governmental *Weibo* accounts, number of government *toutiao* accounts, and number of local regulations, policies, and action plans	12%	Construction of local cyber administration organizations, platforms, institutions, and personnel as well as their capabilities

Assessment items are refined and increased in number. China's Internet Development Index 2017 contained 6 primary indicators, and 16 secondary and 33 tertiary ones. This year, considering that all the regions have put more stress on cyber content building, informatization, and cybersecurity protection, we keep the primary indicators unchanged, but adjust the secondary and tertiary ones and add and refine the assessment items in accordance with the follow-up and collection of data. China's Internet Development Index 2018 is made up of 6 primary indicators and 24 secondary and 51 tertiary ones, covering infrastructure construction, innovation capacity, development of digital economy, Internet application, cyber security, and cyberspace administration. Indicators, aggregate and per capital, qualitative and quantitative, positive and negative, are combined to reflect the Internet development of different regions more comprehensively, objectively, and scientifically.

Weight of the indicators has been adjusted and optimized. For the year 2018, we have adjusted the weight of the primary indicators in China's Internet Development Index. First, the weight of infrastructure construction is reduced to 10%. Network infrastructure is the foundation of Internet development of all provinces, municipalities, and autonomous regions. Thanks to years of construction and development, the fixed and mobile infrastructure of all regions is nearly perfect through constant improvement. Therefore, the weight of this indicator is reduced. Secondly, the weight of innovation capacity has been increased to 20%. Innovation capacity is the inexhaustible driving force for Internet development and the guarantee for permanent competitiveness. To better reflect the input of different regions into technical innovation and talent production, and to encourage and guide all the regions to attach importance to the Internet innovation capacity building, we have increased its weight. Thirdly, we have increased the weight of digital economy, cyber security, and cyber administration. More secondary indicators have been put in the three dimensions in accordance with the Internet development of different regions, including cybersecurity business development, cybersecurity industry, rule of law on the Internet, and cyber administration organization architecture.

Data collection has been expanded and enhanced. This year, we have been following up the Internet development of all regions, sorting the latest progress and representative cases concerned, updating data of all indicators in time, and referring to the statistics of the Internet released by governmental agencies, research institutions, and businesses to ensure that all data are authentic, complete, accurate, and traceable.

2. The Index comprehensively and objectively reflects the general situation of the Internet development of different regions.

According to the latest indicator system and calculation method, we have got the scores of the Internet Development Index of 31 provinces, autonomous regions, and municipalities of China (see Table 2). From the table below, we find that the Internet is mostly developed in economically developed regions like Beijing, Guangdong, Shanghai, and Zhejiang while it is gathering momentum in the middle and western regions.

Guided by General Secretary Xi Jinping's thoughts on cyber power, all provinces, municipalities, and autonomous regions are, in accordance with the unified arrangement by the CPC Central Committee, have achieved great success by enhancing cyber content building, cybersecurity guarantee, and informatization through a series of new creations and practices based on their local reality.

In terms of infrastructure, thanks to the local governments' increase of the fund and policy support, information infrastructure has been reinforced, and the number of Internet users has increased, and the gap between urban and rural areas in information infrastructure is narrowing. In general, the information infrastructure of

Table 2 Top 10 provinces, autonomous regions, and municipalities in the assessment of the internet development

Ranking	Region	Infrastructure construction	Innovation capacity	Development of digital economy	Internet application	Cyber security	Cyberspace administration	Total score
1	Guangdong	7.63	13.30	13.31	22.06	1.06	4.67	62.03
2	Beijing	7.59	13.92	12.43	17.32	2.17	3.00	56.43
3	Shanghai	6.79	8.95	10.42	22.11	1.90	2.55	52.72
4	Zhejiang	6.38	10.06	8.80	21.60	0.79	5.01	52.64
5	Jiangsu	6.55	12.52	9.82	15.97	1.67	4.06	50.59
6	Shandong	6.16	8.64	5.47	16.05	0.39	7.19	43.90
7	Shaanxi	5.41	6.17	7.73	14.74	1.40	3.94	39.39
8	Sichuan	5.82	5.70	5.24	13.54	1.55	5.41	37.26
9	Fujian	5.41	5.37	5.52	15.39	1.27	3.75	36.71
10	Hubei	5.40	6.37	4.90	14.46	1.16	3.42	35.71

economically developed regions like Beijing, Guangdong, and Zhejiang is more developed and regions like Shanxi, Ningxia, and Sichuan are catching up fast in fiber broadband. Chongqing and Hubei are promoting 5G R&D and technical innovation, trying to deploy 5G testing networks. Beijing, Guangdong, and Shanghai are taking the lead in cloud computing infrastructure, laying a foundation for the Internet development of these regions.

In terms of innovation capacity, all regions have begun to take innovation of network information technology as the grip for local innovative development and have been increasing fund and policy support for technical innovation, which has witnessed remarkable results. For example, Beijing and Shanghai, with more talents, take the lead in developing big data and AI, while Shaanxi and Hubei support the R&D of cutting-edge technologies in digital economy by boosting major special projects of science and technology.

In terms of development of digital economy, all regions take e-commerce and the industrial Internet as their development focus, so e-commerce is increasing. Guangdong takes the lead in digital economy by boosting industrial e-commerce and industrial Internet while Zhejiang is facilitating digitalized transformation of traditional industries and rank among the national top in integrated industrial innovation.

In terms of Internet application, new Internet technologies have seen increase in individual, business, governmental, and public service applications, all tending to be diversified, efficient, and convenient. Guangdong, Zhejiang, and Chongqing, by starting with the construction of "digital government", are boosting the governmental information integration while Sichuan, Hubei, and Tianjin are taking the lead in public service application.

In terms of cyber security, cybersecurity awareness and guarantee strength have obviously increased in different regions. Zhejiang and Sichuan have seen the rapid development of the cybersecurity industry through an increase of governmental support, with the scale of their cybersecurity industry and number of relevant businesses taking the lead in China. Anhui and Beijing are among the top regions in cultivation of talents in that field, with great potential in development.

In terms of cyberspace administration, all local governments are, in accordance with the unified requirement of the Central Government, promoting the establishment of provincial and prefectural cyber space administration organizations while formulating laws, policies, and regulations as guidance, and fulfilling local accountability. Zhejiang has launched the initiative of "credit cyber administration" to establish the cyber credit administration mechanism, enrich the cyber governance methods, and improve their effect. Fujian, Tibet, and Qinghai have launched a series of laws and regulations in accordance with the local need to develop the Internet.

IV. Trend of China's Internet Development and Suggestions on Relevant Policies

China is in an era of securing a decisive victory in building a moderately prosperous society in all respects, and the Chinese nation is at a critical juncture in its drive toward rejuvenation. By seizing the historic opportunity brought about by informatization, the country will, by upholding Xi Jinping Thought on Socialism with Chinese Characteristics for a New Era, especially his thoughts on cyber power, keep its foothold in the new historic moment and fulfill its mission in cyber administration. It will speed up the construction for building the country's strength in cyberspace and promote new breakthroughs and leaps in Internet construction, application, and governance.

1. Information infrastructure is playing an increasingly supporting and enabling role in economic and social development, so we should enhance the proactive layout and overall upgrading oriented to future.

Information infrastructure is the strategic public infrastructure in China's economic and social development in the new era. With the iterative IT upgrade and application, Internet of Everything (IoT) and man-machine-thing integration have become reality, with new higher demand for the supporting capability of information infrastructure, which should be immediately improved in domain and capacity and upgraded. The new-generation high-speed, mobile, secure, and ubiquitous information infrastructure should be constructed for networking, digitalization, and intelligent development of economy and society, to provide strong support for the supply-side structural reform and good social order construction. We should upgrade the existing infrastructure's capacity and deployment of emerging technology infrastructure, proceed network structural optimization and critical link dilatancy, and deploy the next-generation Internet in advance while facilitating the overall upgrade to IPv6 to increase traffic unblocking and business-bearing capabilities. Cloud computing centers, CDNs, and IoT and other facilities should be deployed in advance to facilitate comprehensive intelligent information infrastructure which integrates perception, transmission, storage, computing, and processing so that the virtual cyber world and the real world can be integrated in depth. To meet the demand for intelligent manufacturing networking, the construction of new manufacturing foundations like industrial cloud and smart service platforms and industrial Internet should be sped up to form high-ratio, reliant, low-delay, flexible, and fast networks. To meet the demand of social operation and administration, we should transform and upgrade traditional infrastructure to smart power grids, transportation tools, and water affairs facilities, to improve the resource utilization capacity and efficiency.

2. Digital economy's development is coming into the technology-dominating period, so core technology breakthroughs should be made to foster new international competitive advantages.

The global IT technical innovation is witnessing a new-round acceleration and shortened iteration cycle. Technical systems are being restructured fast and emerging technologies such as IoT, cloud computing, big data, and AI are becoming the core driving forces for industrial revolution. A major technical revolution is often a significant opportunity for technical surpassing and "corner overtaking". China will see the critical juncture of high-quality development of its economy, and the innovation and competition advantages generated by "technical development + scale effect" will the solid foundation for the permanent healthy development of the country's economy. The government should, to facilitate the policy and institutional environment for core technology innovation, formulate and implement ground-breaking policies and measures on special tax exemption, loan support, and talent recruitment and cultivation, to encourage State-owned enterprises, private businesses, and colleges to increase their input in the R&D of core technologies. We should accelerate the application of technically innovative achievements to real economy and the popularization of smart factories, digitalized workshops, and production lines, facilitate the overall reform of production factors, industrial chains, technological processes, and business models, and encourage emergence of new business and new models like networking coordination, personalized customization, smart production, and service-oriented manufacturing, to improve the industrial strength and the contribution ratio of technical innovation to economic development. We should enhance the proactive and strategic deployment of standards for basic, cutting-edge, nonsymmetrical, and critical technologies, and accelerate R&D of basic technologies and breakthrough of cutting-edge technologies like AI, quantum computing, quantum communication, and neural network chips to consolidate the technical foundation for the permanent healthy development of China's economy. Businesses should play the major role in accelerating technical R&D and industrial layout to transfer through emerging technologies the advantages in industrial scale and a number of users into advantages for taking the lead and hence new international competitive advantages. A large number of multinationals with strong competitiveness are expected to emerge in China.

3. High-quality content will be the focus of competition among media platforms, so we should speed up restructuring of the order of the cyber content industry and create a healthy cyber ecosystem.

The cyber content industry is witnessing fierce competition. Affected by users' information consumption upgrade and scarcity of quality content resources, quality content has become the new focus of competition among major media platforms, who, therefore, should attach importance to content generation capacity building, introduce positive energy, and innovate the means of cooperating with content generators so that they can jointly plan and generate content products popular among users. They should enhance the profitability of content, foster one-stop service platforms, with quality content as the core, and native advertising and HOBBY as the complementary, and expand the added value of the content. They should optimize content generation and delivery through technical innovation, constantly upgrade technical models and launch new services like personalized products, tailored reading, and targeted-push, and set up the production, checking, and delivery mechanism which is more consistent with content administration. Competent authorities should enhance the intellectual property protection, improve the legal system in the field of network media communication, and reinforce supervision and technical administration over public opinions to upgrade the cyber content industry in healthy competition and make content ecosystem better.

4. Data security is closely related to the security of the nation, industries, and individuals, so we should attach equal importance to security and development and make the best use values of data.

In the information era, data, as a new production factor, is playing an increasingly important role. It is significant basic strategic resource of a nation. China's data will see exponential growth, enjoy extensive application, and influence the development of economy, culture, society, and military affairs. There are closer relations between data security and national security and between social stability and individual interests. How to make the best use of the values of data and guarantee data security is a problem that the Chinese government, businesses, and society have to face and solve. We should attach equal importance to security and development. First, we should take data security as the core of cyber security. Therefore, we should establish the data-centered security protection system, improve rule systems concerning data, clarify the accountability in all steps of data generation and the data ownership and code of conduct concerned, improve data security protection measures, and enhance anti-attack, anti-breach and anti-theft technologies and data security supervision and early warning. Secondly, we should attach more importance to the openness and circulation of data resources and take data flow as the leader of technical, material, capital, and talent flows. Information barriers should be broken and a bigger public data-sharing platform covering the whole country should be formed. This platform should be used in coordination and can be accessed in a unified way to promote the integration and integrated application of data in fields like energy, transportation, and environmental protection. Moreover, we should give full play to the role of data in policy-making and efficiency improvement.

Chapter 1
Accelerated Construction of Information Infrastructure

1.1 Overview

The IT revolution has brought about significant opportunities for economic and social development. China takes cyberspace administration as a priority in the new era, accelerating the construction of information infrastructure and facilitating the economic transformation and upgrading through IT. The 19th National Congress of the Communist Party of China (CPC) proposes developing and building China's strength in cyberspace, making China into a digital and smart country, developing digital economy, cultivating new driving forces, and guaranteeing high-quality development of economy. The Chinese government is boosting the broadband network of increased speed and decreased expenses and universal telecommunication service pilots, and speeding up the deployment of the new-generation information infrastructure to lay a solid foundation for building of the country's strength in cyberspace and for its economic transformation and upgrading.

(1) **Basically overall coverage of fiber broadband**. The coverage rate of broadband in China keeps increasing rapidly. By the end of June 2018, three major telecommunication operators, namely, China Telecom, China Mobile, and China Unicom, saw the number of their fixed broadband users amounting to 378 million, with the household coverage rate of fixed broadband at 78.9%. The country has finished optic fiber access and launched Gbps demonstration. 1.17 billion households are using optical network and 328 million households enjoy optical access, accounting for 87.5% of the total fixed broadband access households. High-speed broadband is being rapidly popularized, and the mainstream broadband access rate has reached 50 Mb/s, with the average download rate close to those top ones of the world.

(2) **Accelerated construction of mobile broadband**. China is promoting its 4G network construction, with dramatic progress in network coverage. 98% of the national population can access 4G network. By the end of June 2018, there were over 3.4 million 4G base stations, the popularization rate of mobile broadband was 90.4%, and the number of 4G users accounted for 73.5% of the total

© Springer Nature Singapore Pte Ltd. 2020
Chinese Academy of Cyberspace Studies, *China Internet Development Report 2018*,
https://doi.org/10.1007/978-981-15-4043-1_1

number of mobile phone users, among the top of the world. Mobile traffic keeps growing rapidly and mobile Internet DOU is rising to over 4 GB. The country is participating in 5G international standard formulation in all aspects, having seen progress in 5G technology R &D tests, constructed the world's largest test outfield and identified 5G spectrum planning and launched the integrated innovation of 5G application in vertical industries.

(3) **Steady layout of the next-generation Internet**. By the end of 2017, the General Office of the CPC Central Committee and the General Office of the State Council published the *Action Plan for Promotion of Large-scale IPv6 Deployment* and the departments concerned thus began to promote large-scale IPv6 deployment through coordination. By the end of June 2018, there were 180 million IPv6 users. IPv6 upgrade and renovation concerning networks, terminals, and operation-supporting systems are witnessing rapid progress, and there is an IPv6 network foundation, but the percentage of networks and APPs supporting IPv6 is to be improved.

(4) **Overall improvement in international communication infrastructure**. International communication transmission networks and overseas network PoPs are basically all over the world. China has 119 PoPs in 32 countries/regions of six continents. There are eight international submarine cables connecting the country with others, with a total capacity of 29.1 Tb/s and a total utilization rate of 47%. It has established 44 cross-border land fiber-optic systems connecting it with 12 neighboring countries, with a total capacity of 38.1 Tb/s and a total utilization rate of 43.2%.

(5) **Universal telecommunication service facilitating poverty alleviation**. By the end of June 2018, the popularization rate of broadband in administrative villages of China had amounted to 97.4%, that of optical networks there to 96%, that of 4G networks there to 95%,and that of broadband in poverty-stricken villages to 94%. Rural e-commerce, smart agriculture, Internet + education, and Internet + medical care are being promoted rapidly. Poverty alleviation through the Internet is a strong backup force for rural rejuvenation and poverty alleviation.

(6) **Steady growth of basic Internet resources**. By the end of August 2018, China had distributed 340 million blocks of IPv4 addresses, accounting for 9.3% of the global total and ranking second of the world; the country had distributed about 27,703 blocks(/32) of IPv6 addresses, ranking second of the world. By the end of March 2018, there were 47.655 million registered domain names, with the number of those with .CN increasing steadily, and domain names with .CN and .COM taking the lead.

1.2 Broadband Network

1.2.1 Fast Popularization of High-Speed Broadband

1. **Project of Increased Speed and Decreased Expenses being carried out and the popularization of broadband service being accelerated**

The broadband network, as a basic, strategic, and guiding facility, plays an important role in accelerating the industrial transformation and upgrading and in promoting digital economy development. The popularization rate of fixed broadband is rising dramatically, almost close to the top of the world. By the end of June 2018, three major telecommunication operators, namely, China Telecom, China Mobile, and China Unicom, saw the number of their fixed broadband users amounting to 378 million, with the household coverage rate of fixed broadband at 78.9%.

Increase of the number of fixed broadband users in China from 2014 to 2018 is shown in Fig. 1.1.

2. **Optical network cities taking shape and the demonstration of Gbps being carried out**

By the end of 2017, China had seen the completion of optical transformation, especially in old neighborhoods, with obvious effects. By the end of June 2018, three major telecommunication operators, namely, China Telecom, China Mobile, and China Unicom, saw the number of their optical network users amounting to 1,170 million, with the number of optical network terminals close to 470 million. Almost all urban households have access to optical networks, and the popularization rate of optical networks in administrative villages had amounted to 96%. In some cities,

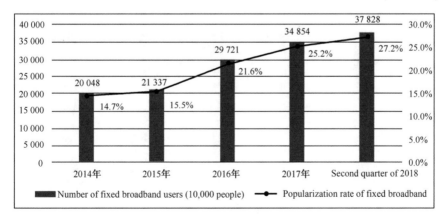

Fig. 1.1 Increase of the number of fixed broadband users in China (2014–2018) (*Source* Ministry of Industry and Information Technology, MIIT)

Gbps network construction has been launched, and the three major telecommunication operators have published their plans on it. In 2017, Gbps demonstration was launched by China Telecom in one hundred cities, covering Shanghai, Beijing, Chengdu, Wuxi, Xiong'an, Guangdong, Anhui, and Yunnan. Shanghai is totally covered by Gbps broadband, having become the world's first city of Gbps.

3. **Optical access as the mainstream and the speed of broadband increasing**

Broadband users are shifting to optical networks. By the end of June 2018, three major telecommunication operators saw the number of their fixed broadband users amounting to 378 million, and 328 million households enjoyed optical access, accounting for 87.5% of the total fixed broadband access households. The number of xDSL users decreased to 2.57%, and that of LAN users to 9.95%. China is one of the leading countries in optical broadband development, with the penetration rate exceeding that of some developed countries, such as Republic of Korea (76.8%), Japan (76.7%), and the United States (12.6%). Change of the percentage of China's fixed broadband users classified by technology is shown in Fig. 1.2.

Broadband networks see their speed rising and average download ratio going up steadily, with the mainstream broadband access ratio increasing from 8 Mb/s to 50 Mb/s. By the end of 2016, only 40% of the users used broadband products of over 50 Mb/s. By the end of June 2018, the number of users of broadband products of over 50 Mb/s and 100 Mb/s was, respectively, 300 million and 200 million, accounting for 80.5 and 53.3%. China's broadband users are witnessing a high speed of access. The change of percentage of users of fixed broadband is shown in Fig. 1.3.

The average download ratio of China's fixed broadband network is close to the top of the world. According to *Report on China's Broadband Ratio* issued in the second quarter by the Broadband Alliance, the average download ratio of China's fixed broadband network is 21.31 Mb/s, the average first screen presentation of fixed

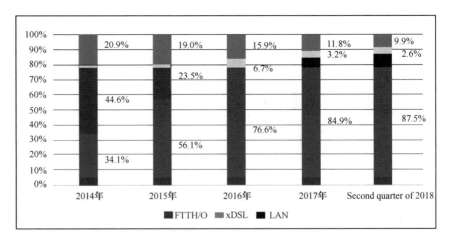

Fig. 1.2 Change of the percentage of China's fixed broadband users classified by technology (*Source* Ministry of Industry and Information Technology)

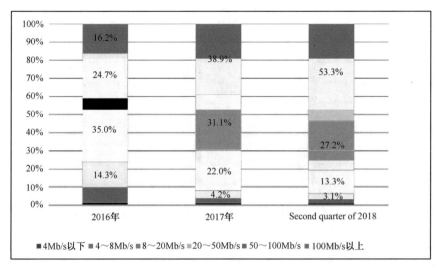

Fig. 1.3 Change of percentage of users of fixed broadband (*Source* Ministry of Industry and Information Technology)

broadband users' browsing the websites is 0.98 s, and the average download ratio of online videos is 18.81 Mb/s, equal to that in advanced countries according to the Akamai report of Internet connection.

1.2.2 Rapid Development of Mobile Broadband

1. **Popularization of 4G networks**

China is boosting 4G network construction, which covers over 98% of its population. By the end of June 2018, there were over 3.4 million 4G base stations, and 1.26 billion users of mobile broadband (3G and 4G), with the popularization rate of 90.4%. By that time, there were 1.11 billion 4G users, accounting for 73.5% of the total number of mobile phone users, with the 4G popularization rate among the top ones of the world. Influenced by the sustaining growth of the number of 4G users, steady growth of traffic cost, and popularization of all kinds of APPs, mobile traffic consumption will maintain high growth and DOU will rise steadily. In June 2018, DOU exceeded 4 GB, with a year-on-year growth of 173%.

Change of the number of mobile phone users and mobile broadband popularization rate in China from 2013 to the second quarter of 2018 is shown in Fig. 1.4.

2. **Promotion of 5G technology**

The standardization of 5G technology of China is witnessing initial success, including breakthroughs in large-scale antennas, advanced encoding, new multiple access, and

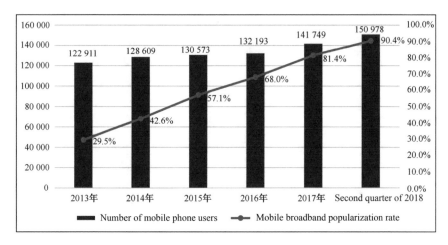

Fig. 1.4 Change of the number of mobile phone users and mobile broadband popularization rate in China from 2013 to the second quarter of 2018 (*Source* Ministry of Industry and Information Technology)

network infrastructure. The country is coordinating in the formulation of 5G standards of foreign industries in the framework of international standard organization of 3GPP (The 3rd Generation Partnership Project)and International Telecommunication Union (ITU), studying and facilitating more innovative technologies being integrated into international standards, and promoting the formation of global 5G standards. Meanwhile, it is accelerating the R&D tests, having constructed the world's biggest 5G testing ground, where the second-stage technical proposal verification and the third-stage testing norm formulation have been completed and the construction of testing environment has been launched. As for spectrum planning, the 5G target frequency band has been clarified and the frequency plan is being formulated; frequency below 6 GHz has been applied to technical R&D testing, and 5G application proposals are being collected openly for two-target frequency band above 6 GHz.

3. **Telecommunication operators promoting 5G planning**

By making full use of the leading role of industrial chains, telecommunication operators have increased input for the industrial development. China Telecom issued the *White Paper of China Telecom's 5G Innovation Demonstration Networks* to start the commercial use of 5G technology in 2020; it clarified the 5G network construction planning in 2018, first oriented to eMBB, with the prior adoption of SA supporting full-scale, full-frequency, and full-networking operation. In 2017, China Mobile succeeded in the world's first end-to-end 5G NR interconnection conforming to the 3GPP R15 standard. It published in 2018 5G commercial products and testing results, as well as the end-to-end technical requirements for 5G scale tests, and launched the 5G Terminal Pioneer Program with 20 terminal industry partners from the world. China Telecom completed in 2017 the deployment of key technologies of the 5G end-to-end network architecture and the 5G Open Lab construction, and it published in 2018 the *White Paper of China Telecom's Edge-Cloud Platform Architecture and Industrial Ecosystem* while launching the large-scale Edge-Cloud pilot projects. The three telecommunication operators of China have planned to launch pilot projects in 12 cities in 2018.

4. **Equipment and terminal businesses' 5G market planning**

Huawei, the biggest telecommunication equipment provider worldwide, takes the lead in 5G standardization and has made breakthroughs in key technology and opened up a number of pre-commercial 5G networks through cooperation with 42 operators from the world. Its Polar Code protocol has been listed as one of the global 5G standards. In early 2018, it officially launched Huawei 5G CPE, the first 5G commercial chip and terminal conforming to 3GPP standard, making itself the first business of the world providing end-to-end 5G solutions. ZTE Corporation has obtained over 1,500 patents and its technical strength has been recognized by operators both at home and abroad. Its pioneering Pre-5G products have been deployed in over 60 networks of over 40 countries. In April 2018, through cooperation with China Mobile, it successfully made the first 5G phone call meeting 3GPP R15 standard, and opened up large-scale outfield sites of end-to-end 5G commercial systems.

1.2.3 Start-up of Large-Scale IPv6 Commercial Use

1. **Launching of the plan for large-scale IPv6 commercial deployment**

In November 2017, the General Office of the CPC Central Committee and the General Office of the State Council jointly issued *Action Plan for Promotion of Large-scale IPv6 Deployment*, which has been implemented by relevant departments, who, in turn, have issued corresponding regulations and have been promoting large-scale IPv6 deployment. Provinces, autonomous regions, and municipalities directly under the Central Government, such as Fujian, Sichuan, Hebei, Tianjin, Guangdong, Xinjiang, Shandong, and Liaoning, have been the first to launch implementation measures for the implementation of the action plan.

2. **Basic industrial capacities**

Network facilities have the capacities for the opening-up of the end-to-end IPv6. The telecommunication operators are speeding up IPv6 upgrading and renovation of backbone networks, metropolitan area networks, access networks, fixed terminals and operation-supporting systems, and new network facilities can almost all support IPv4/IPv6. China Telecom has completely launched IPv6, China Mobile has completed all IPv6 upgrading and renovation, and China Unicom's all backbone Internet facilities can support IPv6. China Telecom has renovated 96% of its metropolitan area networks, China Mobile has initiated IPv6 function in 10 provinces (and municipalities directly under the Central Government) like Beijing, and China Unicom's metropolitan area networks' IPv6 support rate has reached 97%, with those in Beijing, Shanghai, Guangzhou, Shenzhen, Jinan, and Qingdao able to support IPv6 of fixed broadband networks. As for LTE(Long-Term Evolution)end-to-end visit, China Telecom has opened up the mobile networks in 156 cities and distributed IPv6 addresses to LTE users; China Mobile has finished LTE end-to-end renovation in 142 cities in 24 provinces (and municipalities) such as Beijing and Hebei, so that users can access to LTE IPv6; China Unicom has finished IPv6 upgrading of LTE networks, providing LTE network IPv6 data services in some cities (or districts) of 15 provinces (and municipalities). As for interconnection, telecommunication operators have opened up IPv6 inter-network connection broadband at five backbone direct connection points, namely, Beijing, Shanghai, Guangzhou, Zhengzhou, and Chengdu, with the total speed of 3.5 Tb/s.

1.2.4 Continuously Optimized International Network Layout

1. Increasing gateway bandwidth of Internet

From 2017 to 2018, China's gateway bandwidth of the Internet saw a great leap. By the end of June 2018, it had reached 7.28 Tb/s, more than four times that of the same time in 2013, but the gateway bandwidth per capita remains low. By the end of June 2018, it was only 0.02 Mb/s.

Figure 1.5 shows the increase of the gateway bandwidth of China's Internet from the second quarter of 2013 to the second quarter of 2018.

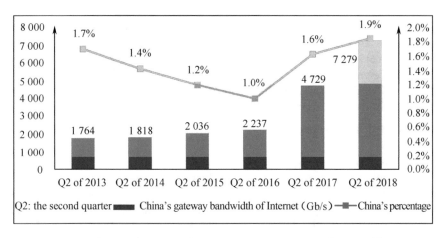

Fig. 1.5 Increase of the gateway bandwidth of China's Internet (Q2 of 2013–Q2 of 2018) (*Source* Ministry of Industry and Information Technology)

2. **Steady overall layout of international business gateways**

China has started to lay out international business gateways, which cover all directions. In Beijing, Shanghai, and Guangzhou there are full-service international communication gateways for voice service, special telephone lines, and Internet oriented to the world; in Urumchi, Kunming, Nanning, Harbin, and Hohhot there are regional and international communication gateways for voice service and special telephone lines oriented to Middle Asia, West Asia, South Asia, Southeast Asia, Mongolia, and the Far East; in Beijing, Shanghai, Guangzhou, Urumchi, Kunming, and Hohhot, there are Internet transfer points serving the neighboring countries and regions. With unified top architecture design, three major telecommunication operators of China have set up more regional gateways for international communication along the coast, which has helped to expand gateway business and overseas grooming spheres.

3. **Overseas network interconnection primarily covering the whole world**

Overseas PoPs are the extension of China's communication networks, contributing to the overseas practice and grooming of international business. By the end of June 2018, the country had laid out 119 PoPs in 32 countries/regions from Asia, Africa, Europe, Americas, and Oceania, with the most in Asia, accounting for 50% of the total, mostly distributed in Southeast Asia and China's Hong Kong, Macau, and Taiwan. Besides Asia, North America, and Europe also see many overseas PoPs of China. But South America, Africa, and Oceania have fewer such PoPs.

4. **International transmission circuits primarily connecting the whole world**

China's international transmission network is made up of international submarine cables and cross-border land cables, which are connected with domestic networks through international communication channel gateways. As for international submarine cable construction, by the end of June 2018, the three major telecommunication operators had a total capacity of 29.1 Tb/s, with the general use rate of 47%. They have joined the construction of international submarine cables landing in Hong Kong, such as AAE-1 and the new Hong Kong-US Cable as well as the construction of other non-landing submarine cables like SMW 5, to expand their international circuit connection and supporting capacity. China has set up cross-border land cable construction systems with 12 neighboring countries, including Russia, Kazakhstan, Pakistan, Vietnam, and Burma. By the end of June 2018, there were 44 such cable systems, with the capacity of 38.1 Tb/s and the overall use rate of 43.2%.

1.2.5 Breakthroughs Made in Air Base Station Technology

1. Tiantong I-01 Satellite: Breakthrough in automatic mobile communication satellite

August 2016 saw the successful launch of the Tiantong I-01 Satellite at the S-waveband of geo-stationary orbit, marking the formal operation of China's automatic mobile communication satellite system. That satellite covers China and its neighboring regions, the Middle East and Africa, and parts around the Pacific and Indian oceans. It can provide reliable and stable around-the-clock mobile communication service irrespective of the weather for individuals, marine transportation, offshore fishing, aviation rescue, travel and scientific investigation, and support voice, messages, and data transmission. It is used in commerce in some provinces like Qinghai, open to emergency communication, field operation, and field and ocean IoT, providing voice and message service.

2. Floating-platform communication technology playing an important role in emergency relief

Floating-platform communication technology, as a supplement of ground network technology, has been developed steadily, having seen breakthroughs in emergency and airborne communication. At present, its R&D and application are seen at small base stations for communication in emergency. Upper air base stations for tethered drone emergency communication are ready for commercial use. China Mobile has tested the upper air base stations for tethered drone emergency communication in Hunan, Beijing, and Inner Mongolia, where they can provide around-the-clock VoLTE(Voice over Long-Term Evolution)and data services for disaster-stricken areas, with its instant messaging covering a range of 50 square kilometers and 5,400 mobile phone users. With strong capability of resisting strong wind and sand, the system can fulfill its mission in an extremely unfavorable natural environment, providing quick and convenient ways for restoring communication in disaster-stricken areas. In the Magnitude −7.0 earthquake of Jiuzhaigou in Sichuan Province and floods in the south in 2017, the unmanned upper air base station of China Mobile provided instant call and Internet access services around the clock for local mobile phone users, meeting the emergency rescue staff's and disaster-victims' demand for communication.

1.2.6 Promotion of Universal Telecommunication Service

1. Deepening of pilot work in universal telecommunication service

China has been carrying out universal telecommunication service, supporting the construction and upgrade of fiber broadband networks in rural and remote areas, facilitating rejuvenation of rural areas and poverty alleviation, and boosting the coordinated development of both urban and rural areas. By the end of 2017, the Ministry of Industry and Information Technology had cooperated with the Ministry of Finance in launching the third group of universal telecommunication pilots, supporting the FTTH (Fiber To The Home) construction and upgrade in 130,000 administrative villages of 28 provinces (and autonomous regions and municipalities), expanding the fiber coverage to villages inaccessible to broadband and contributing to upgrade of fiber in administrative villages whose broadband accessibility was lower than 12 Gb/s. In May 2018, to implement the strategy of rejuvenating rural areas, the Ministry of Finance and the Ministry of Industry and Information Technology continued the pilot work in universal telecommunication service by launching the fourth group of pilots. They are speeding up the 4G network construction in rural and remote areas, giving priority to remote areas such as remote administrative villages, key frontier areas, and islands. The coverage rate of 4G networks in key areas has increased, so that telecommunication service can be balanced and sufficiently developed.

2.   **Improvement of broadband infrastructure capacity in rural areas**

Since the beginning of the universal telecommunication service, the construction of broadband networks in rural areas has seen great achievements, with the capacity and coverage rate increased and household coverage expanded. By the end of June 2018, the popularization rate of broadband in administrative villages of China has amounted to 97.4%, optical fiber access rate of those villages to 96%, and 4G access rate there to 95%. In the administrative villages of Beijing, Tianjin, Shanghai, Jiangsu, Anhui, Henan, Guangdong, and Chongqing, the optical fiber access rate and 4G access rate have reached 100%. It is estimated that by the end of 2018, when the three groups of universal telecommunication pilot projects are completed, over 98% of administrative villages will enjoy optical fiber access, with the capacity amounting to 100 Mb/s.

Administrative villages' optical fiber and 4G access from 2016 to 2018 are shown in Fig. 1.6.

3.   **Network infrastructure contributing to poverty alleviation**

Information infrastructure in poor areas of the countryside keeps being improved. To implement the policies of the CPC Central Committee and the State Council on poverty alleviation, Office of the Central Leading Group for Cyberspace Affairs, National Development and Reform Commission, the State Council Leading Group Office of Poverty Alleviation and Development, and Ministry of Industry and Information Technology jointly issued *Main Points for Poverty Alleviation through Network in 2018*. Network coverage has been expanded, and universal telecommunication pilot projects are implemented; e-commerce is promoted in rural areas and pilot projects of that respect are being carried out; education improvement through the Internet is facilitated and Education Informationization 2.0 program is initiated; information service systems are established for poverty alleviation through

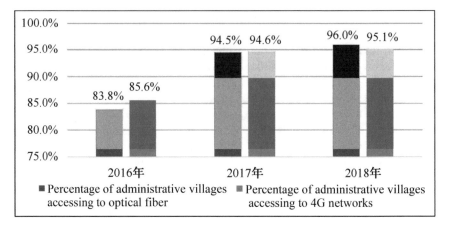

Fig. 1.6 Administrative Villages' optical fiber and 4G access (2016–2018) (*Source* Ministry of Industry and Information Technology)

network infrastructure, and household access to information is facilitated; online charity is implemented, Internet aid for poverty alleviation and education is offered and more Internet businesses are encouraged to participate in poverty alleviation through network infrastructure. In May 2018, the Ministry of Industry and Information Technology issued the *Implementation Plan for Poverty Alleviation through Network* (2018–2020), according to which, priority should be given to the extremely poor areas and designated counties and districts, where the construction of network infrastructure will be boosted and the fiber broadband network extension and 4G coverage will be accelerated. It is expected that by 2020, the broadband network coverage rate in 122, 900 registered poverty-stricken villages will be over 98%. The first three groups of pilot projects of universal telecommunication service have supported 45,000 poverty-stricken villages in fiber network construction. By the end of June 2018, the broadband network access rate in poverty-stricken villages had reached 94%. Digital education resources now cover all schools, which compensates for the lack of teachers for four million students in remote and poverty-stricken areas. Overall coverage of long-distance medical care is promoted. Throughout the country, long-distance medical care coordination networks cover 13,000 medical institutions of all prefectural cities and 1,808 counties. Priority is given to all poverty-stricken counties, so that good medical resources can be distributed down to poverty-stricken counties.

Information application in rural and remote areas is improved. China integrates universal communication service and targeted poverty reduction and eradication, while promoting applications like e-commerce, smart agriculture, Internet + education, and Internet + medical care in rural areas, so that the Internet plays a key role in poverty alleviation. With the improvement of coverage of broadband networks in rural and remote areas, Internet applications like e-commerce are being popularized. Agricultural products are being sold on the Internet, which has helped to increase the income of farmers. APPs like Yinongshe, Grid E-link, and Nongjibao have facilitated the intelligent and grid development of agriculture and hence the agricultural modernization. Meanwhile, APPs are being used in tourism, public service, medical care, and education in rural areas, so information service is being integrated into rural life.

1.3 Application Facilities

1.3.1 Rise of Data Centers and Cloud Computing Platforms

Data centers and cloud computing platforms serve the development of China's digital economy. They are important carriers and nuclear basic facilities supporting the new-generation IT layout. With the implementation of the "Internet+" strategies, they are playing an increasingly important role in cultivating new modes and new business forms, facilitating entrepreneurship and innovation, and promoting industrial structure adjustment. Their construction is generally witnessing an increase.

1. **Cooperation in improving the policy environment of data centers**

(1) **Continuing optimization of data center resources layout**

In accordance with the industrial features of data centers, China categorizes data center construction areas into four, encouraging the construction of large and ultra-large data centers. Especially, the location of data centers for disaster relief applications, with little demand for real-time performance, should be based on climate and energy supply. Guided by policies, energy-intensive regions with a mild climate like Guizhou, Inner Mongolia, and Ningxia have attracted data centers to be located there, which has helped to optimize the layout of all data centers throughout the country, and to solve the problems like the intensity of such centers in the east part and the energy and safety problems resulting from the intensity. With the changes in market, technology, and business, data centers in the west are not used so frequently. In view of that, the Ministry of Industry and Information Technology has compiled the *Development Plan for Internet Data Centers* (IDC), to guide the layout of data center resources.

(2) **Promotion of businesses' use of cloud service and acceleration of construction of cloud platforms for public service**

Three-year Action Plan for Cloud Computing Development (2017–2019) was issued by the Ministry of Industry and Information Technology in March 2017. The plan encourages competent local authorities to construct cloud platforms for public service through cooperation with backbone businesses of cloud computing, and guides software businesses in developing all kinds of SaaS(Software-as-a Service)applications

Fig. 1.7 Distribution of
China's Internet Users and
Websites

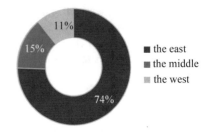

to accelerate the transfer of information systems to cloud platforms, with industrial cloud and government cloud as the pointcuts. In November 2017, the State
Council issued *Guiding Opinions on Enhancing the Industrial Internet through
"Internet Advanced Manufacturing"*, which proposes the goals for businesses' use
of cloud service. In July 2018, the Ministry of Industry and Information Technology
issued *Guidelines for Businesses' Use of Cloud Service (2018–2020)*, which proposes
guiding opinions on paths to cloud service, policy guarantee, and supporting services,
and defines the overall plan for businesses' use of cloud service.

3. **Accelerated data center resource supply in China**

With the rapid development of strategic industries like cloud computing, big data, and
IoT, and the implementation of the strategies like the integration of informatization
and industrialization and "Internet+", the construction of China's data centers is being
accelerated. Local governments and businesses take an active part in the construction,
and the IDC resource supply is increasing steadily. By the end of 2017, 1.1million
IDC racks had been constructed throughout the country, with a year-on-year growth
of 25%. Among all the racks, 65% had been built by basic telecommunication operators, totally 715,000; and 35% had been built by third-party IDC businesses, totally
385,000.

4. **The east area as the gathering place of data center resources**

The large number of clients with a high level of informatization in the east area of
China requires a large number of data centers. There are far more Internet users
in that area than in the west (see Fig. 1.7), with the most in Beijing, Shandong,
Zhejiang, and Jiangsu. Meanwhile, the east area is far more informatized. By the end
of December of 2017, the number of Internet websites on the Chinese Mainland had
reached 4.74 million,[1] with 74% of them in the east. Despite the high cost of data
center construction and operation in eastern cities, where the land and energy cost
remains high, there is increasing investment and demand for data centers there.

---

[1] Source: CNNIC's *41st Statistical Report on Internet Development.*

Fig. 1.8 Layout of China Telecom's data centers in the world (*Source* China Telecom)

5. **Key businesses' acceleration of global layout of data centers and cloud service resources**

China Telecom was the first to have launched the global layout, with over 700 data centers across the world. It was the first telecommunication operator that had "gone out" of China, having established interconnection through the Internet with over 100 advanced Internet operators, with over 400 data centers in China and 13 in overseas as well as over 300 cooperative computer rooms for data centers. It cooperates with over 80 operators throughout the world, with data centers distributed in Asia and the Pacific Region, the United States, and EMEA (Europe, the Middle East, and Africa), as is shown in Fig. 1.8. Alibaba Cloud has data centers and cloud computing platforms across the globe (see Table 1.1), with 11 regional nodes and 44 available zones (AZ). With its independently developed Apsara operating system, Alibaba Cloud provides computing support for one billion users across the world.

1.3.2 Development of CDN Industry in Competition

1. **New technologies and new applications driving the fast growth of the CDN market**

With the development of traditional Internet and mobile Internet, as well as the dramatic increase of the number of multimedia contents and mobile APPs, Internet users' online period and use rate keep increasing, new applications represented by

Table 1.1 Layout of Alibaba cloud's cloud computing platforms in the world

Region	Subregion	Number of AZs	Year of launch
North China	Beijing	6	2013
	Qingdao	2	2012
	Zhangjiakou	2	2014
	Hohhot	1	2017
East China	Shanghai	5	2015
	Hangzhou	7	2011
South China	Shenzhen	4	2014
Hong Kong	Hong Kong	2	2014
Southeast Asia	Singapore	3	2015
	Sidney	1	2016
	Kuala Lumpur	1	2017
	Djakarta	1	2018
South of Asia and the Pacific	Bombay	1	2018
Northeast of Asia and the Pacific	Tokyo	1	2016
East of the United States	Virginia	2	2015
West of the United States	Silicon Valley	2	2014
Europe	Frankfurt	2	2016
The Middle East	Dubai	1	2016

Source Synergy Research Group

online videos like VR, live streaming, and short video boost the dramatic increase of the Internet traffic, and new business forms like IoT and industrial Internet have also potentially promoted the expansion of the CDN market. The integrated development of new technologies like CDN and cloud computing, SDN(Software- Defined Network)and NFV(Network Function Virtualization)has boosted the efficiency of online resources management and utilization and improved the Internet users' experience and the growth of the CDN market. According to CCID statistics, in 2017, the CDN market volume was 13.61 billion *yuan*, with a year-on-year growth of 29.1%. It is expected to reach 2.5 billion *yuan,* with a compound annual growth rate of 35%.

2.   **Formation of the healthy CDN market order**

Since the launching of the "Internet+" program and the "broadband China" initiative, China has issued a series of policies and measures to facilitate the CDN industrial development, and improve the capacity, coverage, and intelligent development of CDN, enhance the bearing and distribution capacity of the Internet traffic, and boost the coordinated development of application facilities like CDN and broadband networks. The optimization of the policy environment contributes to the fast development of China's CDN industry. *Telecommunication Catalog (2015)* lists CDN as an independent form of the Internet business and clarifies the significance and relevant measures of CDN as one of the first classes of value-added business. By the end of August 2018, the Ministry of Industry and Information Technology had issued to 181 operators the permits for value-added telecommunication with CDN.

1.3.3 Large-Scale Layout of Narrow-Band IoT Facilities

NB-IoT technology provides strong basic facilities supporting capacity for China's IoT development. In June 2017, the Ministry of Industry and Information Technology issued *Notice of Promoting NB-IoT Construction and Development by the Office of Ministry of Industry and Information Technology*. Guided by the policies, telecommunication operators, and equipment manufacturers have gathered their strength in developing the IoT, and they take the lead in the large-scale layout of IoT base stations. By the end of 2017, cellular IoT of China had connected 483 million users and the number of NB-IoT base stations had amounted to nearly 570,000. NB-IoT application demonstration is being accelerated. Access management was launched for NB-IoB in China in 2018. By the end of June 2018, 29 permits for NB-IoT equipment access had been issued, all for NB-IoT wireless data terminals.

In January 2017, China Telecom issued NB-IoT V1.0 based on 3GPP. In the first quarter of 2017, outfield testing for NB-IoT V1.0 was done in 12 cities of seven provinces (and municipalities directly under the Central Government), namely, Guangdong, Jiangsu, Zhejiang, Shanghai, Fujian, Sichuan, and Henan. On May 17th, China Telecom set up the world's first commercial NB-IoT with the widest coverage. The operator has started the testing of eMTC, which, by the end of August 2018, had been put into trial commercial use in over 20 cities. eMTC will be put into comprehensive commercial use in accordance with standards and industrial maturity.

NB-IoT base stations of China's three major telecommunication operators by the end of 2017 are shown in Fig. 1.9.

Fig. 1.9 NB-IoT Base Stations of China's Three Major Telecommunication Operators by the End of 2017 (unit: 10,000)

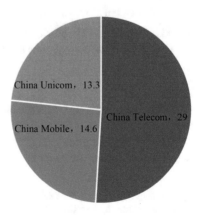

In January 2017, China Mobile set up its first NB-IoT network covering all domains; in the first half of the same year, it carried out large-scale testing of NB-IoT in four cities, namely, Hangzhou, Shanghai, Guangzhou, and Fuzhou, and then expanded the testing to other key cities. China Mobile is dedicated to simultaneously promoting the coordinated development of NB-IoT and eMTC, and the mutual compensation and joint industrial development of technologies. In early 2017, China Mobile launched jointly with Ericsson and Qualcomm the lab testing of end-to-end commercial eMTC products based on 3GPP. So far, small-scope layout and verification of eMTC networks have been done in many cities.

In 2017, China Unicom launched NB-IoT pilot projects in over 10 cities, including Shanghai, Beijing, Guangzhou, and Shenzhen. In September 2017, the Tianjin Office of China Unicom put the first provincial NB-IoT network into full commercial use. China Unicom opened up eMTC testing networks in cities like Beijing in February 2017. It is promoting IoT applications based on eMTC, and will start the commercial layout of eMTC in 2018.

1.3.4 China's Internet Resources Ranking Among the Top in the World

1. **The number of IPv4 addresses ranking second in the world and that of Guangdong, Beijing, and Zhejiang ranking first**

By the end of August 2018, China had distributed about 340 million IPv4 addresses, with an increase of 20 million in comparison with the number in 2017. The total number accounted for 9.3% of the world's total, ranking second in the world, following that of the United States, who has 1,606 million. Therefore, China has a long way to go. The number of IPv4 addresses in Guangdong, Beijing, and Zhejiang ranks first.

China's use of IPv4 addresses is shown in Fig. 1.10.

2. **The number of IPv6 addresses ranking second in the world and the notification rate[2] having increased dramatically in comparison with that of last year**

By the end of August 2018, China had 27,703(/32) IPv6 addresses, with an increase of 29.98% in comparison with the number in 2017. The number ranks second in the world, lower than that (45, 273(/32) of the United States. All regions and departments have been implementing the *Action Plan for Promotion of Large-scale IPv6 Deployment* since it was launched at the end of 2017, with China's IPv6 advertising

---

[2]Advertising rate refers to "% of Advertised space" in APNIC resources reports. It is the ration of total/64 s advertised and total/64 s allocated. The advertising of IP addresses in the network is that of one section of the IP address through the prefix of the address. The advertised address can be routed in the network.

Fig. 1.10 China's Use of
IPv4 Addresses (*Source*
IPIP.net)

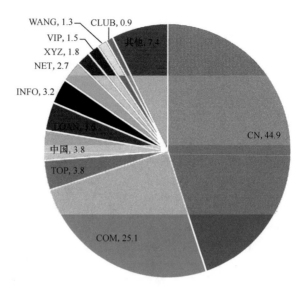

rate increasing dramatically. By the end of August 2018, it was 9.9%, 17 times higher than that of the same period in 2017.

3. **The domain name registration market slowing down and .CN and .COM maintaining their dominant position**

By the end of March 2018, China's domain name registration market volume had decreased by 2% on a year-on-year basis, amounting to 47.655 million. Among the names, .CN has seen a steady increase, with its registered names amounting to approx. 21.4 million. China's top ten registered domain names include the leading one .CN/. 中国, three traditional gTLD names, and five new gTLD names, all accounting for 91.5% of the market share (Fig. 1.11).

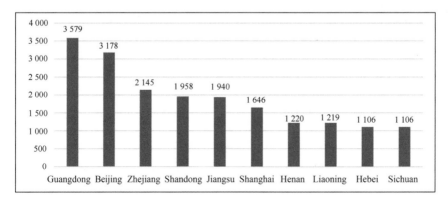

Fig. 1.11 Leading Domain Name Market of China's Top 20 Registered Domain Names (Unit: 10,000) (*Source* CNNIC and ICANN)

Chapter 2
Continuous Network Information Technology Development

2.1 Overview

The year 2018 a saw healthy development of China's network information technology, with AI, blockchain, quantum computing, and brain-like computing catching the eye. The R&D of China's network information technology is mostly done in the application, and the country has a long way to go to catch up with advanced countries in the basic and underlying technology. Generally speaking, its network information technology is developing in the following ways:

(1) China's network information technology generally lags behind that of advanced countries, and it has shortcomings or deficiencies in some aspects. For example, there is little research in neuromorphic computing, and slow development in GPU (Graphic Processing Unit) and DSP (Digital Signal Processing) chips research. RF chip technology is far behind, and storage chips are not massively manufactured yet. There is a two-generation gap between China's chip process technology and that of the advanced technology of the world. Such widely applied industrial software as CAD/CAE nuclear technologies is mostly in the hand of overseas businesses.

(2) China has achieved progress in nonsymmetrical and cutting-edge technologies. For instance, the research staff from the University of Science and Technology of China has been the first in the world that has worked out 18-byte light quantum entanglement. They have independently worked out the new-generation prototype of the supercomputer with exascale computing. They have also been the first to have used the four-dimensional entangled state to obtain quantum dense coding.

(3) Cyber businesses play an increasingly important role in the development of network information technology, and have shifted their research from application theories to basic theories, and from one field to multiple relevant fields, with their input in R&D increasing year by year. Relying on their financial strength, they are uniting higher education and research institutions to carry out

Chinese Academy of Cyberspace Studies, *China Internet Development Report 2018*, https://doi.org/10.1007/978-981-15-4043-1_2

research in network information technology, and put stress on the introduction and cultivation of technical talents.

2.2 Rapid Development of Basic Network Information Technology

2.2.1 Steady Development of Advanced Computing Technology

1. Cloud computing

In 2018, with the quick rising of segmented PaaS areas like container, microservice, and DevOps,[1] Iaas and PaaS are being integrated, facilitating cyber businesses' development, deployment, distribution, and testing application in the environment of cloud, and the integration of networking, computing, and storage. Besides, GPU is developing to cloud transformation, which has helped to reduce the cost of hardware and the specified demand of technology, offering good support for the integrated service of cloud computing and AI. Mainstream cloud computing manufacturers of China are launching GPU cloud host renting, providing accelerated computing for specific scenarios like graphic database, computing finance, earthquake analysis, molecular modeling, and genomics, and hence facilitating corporate innovation.

2. High-performance computing

By June 2018, a national high-performance computing service environment constituted by 17 high-performance computing centers had been established, with the resource capacity listed as one of the top in the world and cutting-edge results produced one after another.[2] At the same time, Sunway Taihu Light was ranked second in the world's top 500 supercomputers, with its floating-point arithmetic speed at 93 petaflops per second. In July 2018, the new-generation supercomputer of Exascale Tianhe No.3 E-class prototype was developed independently by China. Three high-performance computing and communication chips of proprietary intellectual property rights have been adopted for it, whose computing capacity will be increased by 200 times and storage capacity by 100 in comparison with that of Tianhe No. 1.

---

[1]DevOps: It is a combination of the words " Development" and "Operations", a unified term of processes, methods, and systems, which are used for promoting communication, coordination, and integration among development, technology, and quality guarantee sectors.

[2]Zheng Xiaohuan, et al. Analysis of Global High-performance Computing Development [J]. World Sci-Tech R&D.

3. **Neuromorphic computing**

Neuromorphic computing technology will be the next priority in high-performance computing technology. It will contribute to a dramatic increase of data processing and machine learning capacity. China is seeing little research in that area. Peking University has launched some neuromorphic computing application research based on resistance change devices and has seen some achievements in synaptic structure and realization. Some start-up businesses of China have joined in the R&D of neuromorphic computing chips. For instance, Shanghai Westwell launched in July 2016 the Westwell Brain, a real-time simulator of human brain, with 10 billion nerve cells, and the brain-like commercial chips with 50 million nerve cells.

2.2.2 Progress in IC Technology

1. **Computing chips**

China's computing chip industry has developed fast though it was started a little bit later. CPU chips are a series of processor products represented by MIPS, ARM, X86, and POWER. Some businesses have begun their R&D of chips like GPU and DSP, and ASIC chips have competitive strength in some fields.

A number of architectures coexist in the CPU chip industry of China. As for X86, Shanghai Zhaoxin has worked out the complete SoC (System-on-a-Chip) solution based on its own GPU and Chipset. The KX-5000 processors it launched at the end of 2017 are the first Chinese CPU for general use, which can support DDR 4 storage. Haiguang began to carry out R &D of processor CPU with the authorization of AMD, and the chips were put into massive production in July 2018. PKUnity has developed X86-compatible CPU with the authorization of UniCore and AMD and it is also engaged in R&D and sales of supporting software and hardware in application for low and medium-level desktops and high-end built-in computers. As for MIPS, Loongson has been adhering to independent R&D of microstructures and compilers. It has developed six of the former including GS464V and GS464E, and LCC(Local C Compiler)to optimize its microstructure. Ingenic's CPU product is XBust. Besides the adoption of MIPS instruction, it has expanded its SIMD instruction set. As for ARM, Huawei HiSilicon, ZTE SANECHIPS, Spreadtrum, Leadcore Technology, Ingenic, and Allwinner Technology, all have experience in ARM chip R&D, and they have launched CPU products in application like mobile terminal and high-performance computing. As for servers, Tianjin Phytium and Huaxintong Semi-technology have obtained the permit of architecture. Phytium processors have seen the shift from SPARC to ARM 64, and the CPU of FT-1500A and FT-2000 series is being applied in industries. Besides, Alpha has been adopted for Sunway series of processors, oriented for high-performance computing and servers; Zhongsheng Hongxin Information Technology has undertaken the R&D of the whole set of POWER CPU technology, having launched CP1 and CP2 chips, whose R&D is based on POWER 8 chips of IBM. CP3 was defined in mid-2017, and it will be launched around the year 2020.

China has to catch up in GPU chips, with only Jingjia Micro, Zhaoxin, VeriSilicon, and Loongson having made some progress. In September 2018, Jingjia Micro launched JM 7200 chips, for which 28 nm techniques have been adopted, and which can meet the demand of high-end embedding applications and computers; Zhaoxin completed the development of Elite 1000 and Elite 2000. VeriSilicon began its GPU business after it acquired GPU IP graphic chip technology. Some businesses have begun to construct SoC systems by adopting overseas GPU IP. For instance, GPU IP has been adopted for Loongson's 2H and PUnity's Tiandao. Huawei HiSilicon has adopted MALI of ARM.

2. **Memory chips**

China also has to catch up in memory chip production despite the development of its independent core technology, which has, however, failed to see the massive production of the chips. Major businesses of memory chips are YMTC, Fujian JHICC, Hefei Ruili, and GigaDevice.

China has seen the development of DRAM by Fujian JHICC and Hefei Ruili. The former orients its products to niche markets, having begun the trial production of 8 Gb LPDDR 4 memory chips. The latter has been expected to launch 19-nm 8 Gb DDR 4 memory chips. Unigroup Guoxin is mainly engaged in the design and modeling of DRAM. It is carrying out DDR 4 memory chip R&D, which is expected to be completed at the end of 2018. The products will be launched into the market by then.

As for NAND Flash, YMTC, based on the XMC R&D and production capacity in terms of 12in IC technology, succeeded in producing China's first 3D NAND flash chip in 2017 through independent R&D and international cooperation. XtackingTM of 3D NAND has also been launched, which will help to increase the I/O performance and memory density of 3D NAND Flash and to shorten its market cycle. YMTC has applied this technology to the development of second-generation 3D NAND products.

3. **Communication chips**

Domestic businesses' level of design of mobile information terminal baseband chip is the same as that of their counterparts in other countries. LTE (Long-Term Evolution) multimode and multifrequency 64-digit multi-nuclear SoC has seen its technique developed to 7 nm. HiSilicon has made breakthroughs in some single areas, renewing the products every 2 years, ranking first alternatively with Qualcomm in the world. In August 2018, HiSilicon publicized its Kylin 980 mobile phone SoC, with the adoption of the 7 nm technique. The Kylin series of mobile phone chips publicized by HiSilicon are shown in Table 2.1.

China has to catch up in the field of RF. Some domestic businesses are taking the lead in the market of 2G and 3G mobile phone PA, and they are doing R&D in 4G and 5G PA products. But their technology in that respect is relatively behind, having to catch up with Win Semiconductor, Advanced Wireless Semiconductor, Qorvo,

Table 2.1 Kylin series of Mobile Phone chips publicized by HiSilicon

System chips	Kylin 980	Kylin 970	Kylin 960
CPU	2*A76@2.60 GHz 2*A76@1.92 GHz @512 KB L2 4*A55@1.80 GHz @128 KB L2 4 MB DSU L3	4*A73@2.36 GHz 4*A53@1.84 GHz 2 MB L2	4*A73@2.36 GHz 4*A53@1.84 GHz 2 MB L2
GPU	ARM Mali-G76MP10 @720 MHz	ARM Mali-G72MP12 @746 MHz	ARM Mali-G71MP8 @1037 MHz
LPDDR4 storage	4*16位CH LPDDR4X@2133 MHz 34.1 GB/s	4*16位CH LPDDR4X@1833 MHz 29.9 GB/s	4*16位CH LPDDR4@1866 MHz 29.9 GB/s
Memory I/F	UFS 2.1	UFS 2.1	UFS 2.1
ISP/video camera	Speed of new double ISP +46% speed	Double 14-digit ISP	Double 14-digit ISP (improved version)
Coding/decoding	2160p60 decoding 2160p decoding	2160p60 decoding 2160p30 decoding	1080p H.264 Decoding and coding 2160p30 HEVC decoding
System chips	Kylin 980	Kylin 970	Kylin 960
Integrated modem	Kylin 980 integrated LTE (Type 21/18) DL = 1400 Mb/s 4*4 MIMO 3*20 MHz CA, 256-QAM (5CA without MIMO) UL = 200 Mb/s 2*2 MIMO 1*20 MHz CA, 256-QAM	Kylin 970 integrated LTE (Type 18/13) DL = 1200 Mb/s 5*20 MHz CA, 256-QAM UL = 150 Mb/s 2*20 MHz CA, 64-QAM	Kylin 960 integrated LTE (Type 12/13) DL = 600 Mb/s 4*20 MHz CA, 64-QAM UL = 150 Mb/s 2*20 MHz CA, 64-QAM
Sensor center	I8	I7	I6
NPU	Dual@>2*perf	Yes	No
Technique	TSMC 7 nm	TSMC 10 nm	TSMC 16 nm FFC

and Cree, the leading ones. As for SAW filters, some Chinese research institutions and businesses have started relevant R&D and made some progress. For instance, Huaying Electronics and Wuxi HD Shoulder have the capacity of producing SAW filters in large quantities, but they are generally still not so strong.

4. Techniques of chips

On the principle of ensuring the simultaneous development of advanced techniques and characteristic ones, China, by following Moore's Law, is enhancing the R&D

to keep up with the IC advanced processing of the world. The domestic processing technology has witnessed 16 nm/14 nm, two generations behind the world's advanced level. Moreover, the country is boosting characteristic techniques, by constructing production lines of compound semiconductors and MEMS and enhancing the R&D of power management, power devices, and image sensors to meet the demand of design. Leading manufacturers focus on platforms of such processing technologies as NOR Flash, MCU CMOS image sensors, and HV, developing machines compatible with logic processing and allocating capacity properly.

2.2.3   Acceleration of Commercialized Application of Software Technologies

1. **Operating system (OS)**

In terms of IoT OS, Alibaba launched in September 2017 AliOS, a built-in IoT operating system and the cloud-terminal integration IoT platform combining minimalist development, cloud-terminal integration, component enriching, and safety protection, which have been applied in the smart home, the smart city, and new transportation. As for AI OS, Baidu launched in July 2018 DuerOS 3.0, a dialogue AI OS, which allows equipment monitoring, information query, and link service in an interactive way of natural language dialogues. It has been applied to mobile phones, robots, and wearables. Meanwhile, Baidu launched an open platform for a voice AI ecosystem to support the access of the third-party developer. As for cloud OS, container service based on Google Kubernetes like Tencent TKE and Baidu CCE container search engines are the major engines. In the first half of the year 2018, Alibaba Cloud launched its Pouch Container, which provides rich containers, strong isolation, and P2P distributed memory, with kernel, standard and Kubernetes adaptations.

2. **Cloud storage**

The contribution of commercial practice with Chinese characteristics to open-source community is showing itself. In March 2018, Tencent launched blueking CMDB 2.0.0, with new solutions such as microservice, container, and Golang language. It is a platform with a unified configuration and management for different industries and infrastructure. In May 2018, Alibaba launched DragonFly, a P2P image and file distributed storage system, applicable to PB-level data such as business application distribution, buffer memory, log phone, and image distribution. It is one of Alibaba's internal storage facilities, with its reliability at 99.9999%. In August 2018, Baidu launched BaikalDB, an enhanced distributed structural database based on MySQL protocol. All MySQL applications can be transferred to BaikalDB seamlessly.

3. **Software-defined network (SDN)**

In May 2017, Tencent opened more than one unit in its WeChat application, which allows networking and monitoring of all edge devices and will evolve into a general

SDN solution to edge computing. Alibaba and Baidu can also provide CDN similar to that of Amazon. Generally speaking, China's manufacturers can provide relatively single SDN strategies and facilities, and their solutions within and across data centers can only be tested within themselves, so they have to catch up in developing open-source products that can be put into large-scale commercial use.

2.3 Emerging Cutting-Edge IT Highlights

2.3.1 Attractive AI Applications

1. AI computing

Thanks to the joint effort of universities and research institutes, China has seen some progress in basic AI models. For instance, Zhou Zhihua from Nanjing University has worked out the model named Deep Forest, which, different from the deep neural network, can do representative learning and determine automatically the complexity of the model through the cascade forest structure. It can be competitive with the deep neural network in performance, so it can be a backup solution to the latter in the future.

2. Open software platforms

Influential AI software platforms in China include Baidu's PaddlePaddle deep learning platform and Apollo automatic driving platform, Alibaba's ET City Brain, and Tencent's auxiliary medical diagnosis and treatment platform. Apollo automatic driving platform offers a practical driverless system and hence a systematic solution covering sensing, decision-making, and implementation. With three forms, namely, open code, open capacity, and open data, it enables its developer and eco-partners to cooperate, with the number of the latter amounting to over one hundred. ET City Brain includes computing, data resource, intelligent and application-supporting platforms, covering computing capacity, data algorithm, and management model of urban transportation, medical care, city management, environment, tourism, urban planning, peace, and livelihood service. It can be opened to all eco-participants of city management. The auxiliary medical diagnosis and treatment platform came from the AI auxiliary medical diagnosis and treatment engine of Tencent Miying. The engine can be used through an open access by hospitals and medical information manufacturers, covering intelligent medical services from beginning to end of a diagnosis. The platform contains most of the open authoritative medical repository.

3. AI chips

In recent years, China has witnessed rapid development in AI chips, catching up with the world. Huawei, Baidu, and Alibaba have all begun to manufacture chips, and start-ups like Cambricon, Horizon Robotics, and DeePhi have also launched AI processing chips. In July 2018, Baidu launched its independently developed full-function AI chip named Kunlun to satisfy the demand for training and deduction. Alibaba DAMO Academy (for Discovery, Adventure, Momentum, and Outlook) has announced that it is developing the AI chip Ali NPU, which will be applied to image video analysis and machine learning. Meanwhile, Pingtouge (with the domain name T-head.cn), a semiconductor business, has been founded by Alibaba to promote the cloud-terminal integration of chips. Huawei issued Kylin 980, the new-generation AI mobile chip, at the IFA2018 electronics exhibition in Germany. It is the world's first mobile phone chip adopting 7 nm process technology, with its performance increased by 75% and its efficiency by 58% in comparison with the last-generation processor. Cambricon launched in 2017 Cambricon 1A, the world's first commercial deep learning chip; in 2018 Cambricon 1 M, the third-generation chip whose performance is 10 times higher than that of 1A; and Cambricon MLU 100 cloud AI chip and PCIe interface, a cloud intelligent processing computing card. Horizon Robotics has launched BPU, a self-designed AI intelligent processor architecture, and Sunrise1.0, a built-in AI vision chip, which can process large-scale face detection and follow-up and video structure. DeePhi launched Tingtao series named SoC, carrying 28 nm TSMC process of MediaTek and realizing the peak performance 4.1 TOPS by consuming only 1.1 W of power. Vimicro has developed Starlight Smart No. 1 (VC0758), China's first built-in neural network chip. China's ASIC tailored chip design is at the world's advanced level. According to the fact that ASIC will develop into a key AI chip, the country's AI chip will see a bright prospect.

4. **Basic application technology**

(1) Speech recognition technology (SRT)

Some Chinese businesses like Baidu, Unisound, and Aispeech take the lead in SRT. In January 2018, Baidu launched Deep Peak 2, its latest achievement in speech recognition. It has also launched Duer, a robot that can converse continuously, DuerOS voice platform, voice input, and Xiaodu Zaijia, a smart speaker. In September 2018, Unisound jointly drafted with Ping An Health Cloud *Chinese Speech Recognition Difficulty Level Certification*, according to which the difficulty and capacity of smart Chinese SRT will be classified.

(2) Vision recognition technology

Vision recognition has been applied to safety protection, education, transportation, finance, medical treatment, and smart equipment. Five key businesses of vision recognition are Hikvision, SenseTime, Megvii, YITU, and CloudWalk. For example, in cooperation with Vivo, SenseTime, in 2018, built body-shaping technology based on the SenseAR Platform in the latest-launched Vivo X23, providing convenience for the users of short video. Megvii won first prize by defeating Google DeepMind in the AVA challenge contest of CVPR[3] 2018. In June 2018, YITU and West China Hospital,Sichuan University launched Clinical Lung Cancer Research Database, China's first medical big data AI application based on multidimensional clinical data

---

[3]Conference on Computer Vision and Pattern Recognition.

smart treatment, and the world's first smart multidiscipline lung cancer diagnosis system. CloudWalk has launched 3D optical face recognition technology, which has been applied to FaceID of iPhone X.

(3)   Natural language processing technology

Natural language processing technology has been applied to machine translation, information search, personalized recommendation, and Q&A system. Baidu and ByteDance are developing and applying relevant technologies. Baidu translation is witnessing increasing correction rate thanks to natural language processing technology, with its effect close to the professional level. Besides, Baidu has launched the APP named Simple Search. Toutiao and Douyin can gather a large number of users within a short time and ByteDance has invented a writing robot named Zhang Xiaoming.

2.3.2   Innovative Applications of Emerging Blockchain Technology

China is witnessing continuous innovations in blockchain technology, which has been applied rapidly in supply chain finance, credit, product tracing, copyright transaction, digital ID, and e-evidence. In August 2018, Tencent assisted Shenzhen Municipal Taxation Administration in issuing the first blockchain e-invoice. In January of the same year, Alibaba launched the Ant Blockchain, which is an optimization of the "blockchain + charity" that had been put into trial use before. In March 2018, Alibaba announced that TMall Haitao would, based on blockchain technology, follow up, upload, and check the full-chain logistic information of cross-border imports. According to IPRdaily data, Alibaba has applied for 90 patents, ranking first in the world. On September 26, 2018, Baidu publicized the blockchain white paper and launched the blockchain open-source platform named XuperChain and six APPs developed on the basis of XuperChain, namely, PIC-CHAIN, DUYUZHOU, Baidu Huixue, Treasure Box, Baike on the blockchain, and Hubert.

2.3.3   Quantum Information Technology Developing in Line with that of the World

China has seen much progress in quantum information technology, and takes the lead in quantum communication, but has to catch up with some countries in quantum computing and sensing.

1. **Quantum communication**

In terms of construction of quantum entanglement between super-distance nodes, Pan Jianwei and his team, with the help of Mozi Quantum Experiment Satellite Platform, succeeded in 1,203 km satellite-earth entanglement distribution photon system in July 2018, and in the 1,400 km quantum teleportation experiment in August 2018. With regard to the field quantum key, Pang Jianwei and his team, succeeded also in August 2018, in satellite-earth decoy-state quantum key distribution and in increasing the quantum key distribution distance to 1,200 km, with the help of the Mozi Satellite. In July 2018, Guo Guangcan and his team succeeded in quantum dense coding for the first time by using a four-dimensional entanglement state, increasing the channel capacity to 2.09 B, a new record of the world.

2. **Quantum computing**

Physical realization of quantum computing can be based on systems such as non-topological superconductivity, ion trap, and semiconductor quantum dot, or topological quantum systems of non-mediocre topological nature. Quantum computing technology based on superconductivity is mature. It has been adopted by Alibaba and Origin Quantum. The latter has launched the 6-bit superconductivity quantum chip Spcd-6Q. The 11-bit superconductivity quantum computing service jointly developed by the Chinese Academy of Sciences and Alibaba was launched onto the quantum computing cloud platform in February 2018. It is the world's second system providing over 10 bits of quantum computing cloud service. Quantum computing based on ion trap has seen its experimental parameters being optimized in recent years, in addition to its high quantum gate fidelity. For instance, Jin Qihuan and his team from Tsinghua University increased the single-bit decoherence period to 10 min. With the maturity of the technology, ion trap is expected to be the competitor of superconductivity in the construction of practical quantum computers. Semiconductor quantum dot and photon quantum are not so competitive in comparison with the above-mentioned two systems, but they are more applicable to specific scenarios. For instance, Jin Xianmin and his team from Shanghai Jiao Tong University succeeded in their experiment of quantum walk with the photon quantum chip of 49×49 node.

Origin Quantum and Chinese Academy of Sciences and Alibaba have made world-advanced achievements in quantum simulation. The former succeeded in February 2018 in simulating 64-quantum bit and 22-layer deep random quantum circuit by transferring the original circuit into paralleled sub-circuits through the disintegration of the double-quantum bit logic gate; and the latter succeeded in May 2018s in simulating 81-quantum bit and 40-layer Google random quantum circuit through the Taizhang quantum simulator. This, to some extent, urges people to rethink the minimum quantum bit quantity and computing depth needed by quantum supremacy.

Column 1: Quantum Supremacy: A Comparison Between Classical Algorithm and Computing Power of Quantum

Quantum Supremacy is a term coined by Jhon Preskill from the California Institute of Technology in 2012. It is used to describe some special problems. A proper quantum algorithm can allow super polynomial acceleration in comparison with the best classical algorithm.

Universally recognized problems include factorization, Boson sampling, and random quantum circuit output sampling. What is worth stressing is that people only believe rather than prove that the failure to find any classical algorithm comparable to quantum algorithm is not enough for concluding that there is no such classical algorithm. This is a key reason why some researchers are worried about quantum computing. The latest progress that has impetus on quantum supremacy include the Taizhang simulator's successful simulation of 81-quantum bit and 40-layer random quantum circuit output sampling, and the proposal of the index-accelerated classical algorithm.

However, the excellent performance of quantum computing in the above-mentioned problem-solving can bring benefits in the present condition (before a better classical algorithm is found), including but not limited to medicine molecule simulation and big data processing. Most researchers have a positive attitude toward quantum supremacy because of some evidence (Certainly, there is some slight possibility of being refuted), and all countries are speeding up the deployment of quantum computing. Therefore, it is reasonable and necessary for China to increase input and attract relevant talents in the field of quantum computing.

3.   **Quantum sensing**

Developed countries in Europe and America are taking the lead in quantum sensing. China is trying to catch up and narrow the gap. In August 2018, Guo Guangcan and his team from University of Science and Technology of China and Nanjing University succeeded in single-photon Kerr effect measurement of Heisenberg extreme accuracy by means of optimized weak measurement, showing the advantage of accuracy quantum measurement in practical measurement, and providing a new way of quantum accuracy measurement and weak quantum measurement. In September of the same year, Du Jiangfeng and his team from the University of Science and Technology of China, by using the diamond NV color core quantum sensor, succeeded in detecting the magnetic resonance spectrum of a single protein molecule, which was a breakthrough in single-molecule resonance.

**Column 2: Latest Achievement of Chinese Academy
of Sciences—Alibaba Quantum Computing Laboratory**

Chinese Academy of Sciences–Alibaba Quantum Computing Laboratory, founded in Shanghai in July 2015, aims to do prospective research in the field of quantum information science. It is a representative of the fast development of that field in China, coming close to Google, IBM, and Intel in quantum computing.

On March 29, 2017, Alibaba Cloud announced the case of cloud quantum encryption communication at the Computing Conference Shenzhen Summit and hence became the first cloud computing company of the world in providing unconditional safe data transmission service through multiple quantum secure transport domain.

On May 3, 2017, University of Science and Technology of China, Chinese Academy of Sciences–Alibaba Quantum Computing Laboratory, Zhejiang University, and Institute of Physics of Chinese Academy of Sciences jointly launched the photon quantum computer prototype, with its Boson sampling 24,000 times faster than others of its kind.

On October 11, 2017, Alibaba Cloud announced the launching of the Quantum Computing Cloud Platform at the Computing Conference (Hangzhou), together with the Institute of Quantum Information and Quantum Science and Technology Innovation (Shanghai) of Chinese Academy of Sciences.

On February 22, 2018, 11-quantum bit superconductivity quantum computing service was launched onto the quantum computing cloud platform, which is an open channel for the public to experience, know about, and study quantum computing, as the second system after IBM providing over 10-quantum bit quantum computing cloud service.

On May 8, 2018, Shi Yaoyun and his team from Alibaba Quantum Laboratory announced that they had developed the world's strongest simulating quantum platform Taizhang, which, relying on Alibaba Cloud resources, has succeeded in simulating 81-quantum bit and 40-layer random quantum circuit, posing a challenge to Google researcher's argumentation on 50-quantum bit and 40-layer quantum supremacy and urging people to rethink the comparison of computing power between classical computers and quantum computers.

2.4   Network Information Technology Bringing About Revolution in Traditional Areas

2.4.1   Industrial Internet Facilitating Intelligent Manufacturing

1.   Accelerating industrial Internet platform construction

On February 27, 2018, China's first national-level industrial Internet demonstration platform Haier COSMOPlat was approved. It is providing assistance for 320 million users of 35,000 businesses in accelerating product innovation and intelligent manufacturing. CASICloud has been connected to industrial cloud platforms in seven provinces (and municipalities). The number of registered businesses is close to 1.1 million, with micro, small, and medium-sized businesses accounting for over 90%, and over 10,000 machines connected with the platforms. ROOTCLOUD of SANY HEAVY INDUSTRY CO., LTD. has been connected with over 400,000 sets of energy equipment, textile equipment, port machinery, and engineering machinery. Nearly 10,000 parameters have been collected, with the value of connected assets amounting to hundreds of billions of *yuan*. Despite its rapid development, China is behind developed countries in some key areas of industrial cloud. For instance, the country has a weak foundation for industrial control system/high-end industrial software, little core technology of platform data collection/development tools/APP service, little security protection, and imperfect standard systems.

2.   Increasing effect of AI application

AI has witnessed the shift of its R&D to its commercialized use, with an increasing number and expansion of its applications. AI platforms for supply, research, production, and sales are being promoted. In August of 2018, Alibaba Cloud launched its ET industrial brain platform, supporting its partners in the integration of industrial know-how, big data capacity and AI algorithm, tailor-making AI APPs for manufacturers, providing industrial businesses with fast cloud access of production data and training exclusive AI APPs for manufacturers. A key breakthrough of China's AI businesses is failure prediction and intelligent maintenance in manufacturing. For instance, Shenzhen Xuanyu Tech predicts the time of high-end CNC machine tool knife replacement through machine learning, which helps to shorten the production lines' break for knife replacement from scores of minutes to only several minutes. Such technology has been applied to Foxconn's iPhone 8 production line. The alarming model developed by WSYEngine can process historical data through machine learning, and based on real-time sensor data, predict and inform the staff of the replacement of parts that are to fail.

3.   Breakthroughs in China's industrial OS

Industrial OS is the "soul" of industrial auto-control, and the pillar of Industry 4.0. Apsara developed by Alibaba Cloud provides a good basic facility solution for industrial OS, promoting it to develop toward cloud OS. Ningbo Industrial Internet Institute launched in April 2018 "supOS", which is an integration of the unified configuration development platform of industrial intelligent APPs, big data analysis platform, industrial AI engine service, and industrial intelligent APPs. It can be applied to process industries such as petrochemical, nuclear electricity, medicine, and oil and gas production. Shenyang Machine Tool Group launched at the end of 2017 the world's first industry-oriented mobile control OS named i5OS. Across the globe, MindSphere (an open IoT OS based on a cloud developed by Siemens), ABB Ability (an innovative cloud computing platform developed by ABB for monitoring, optimizing and controlling electric system), IIOT (industrial IoT) are the leaders of industrial OS. The launching of China's industrial OS with proprietary intellectual property rights is the starting point breaking the above-mentioned leaders' monopoly, marking the key breakthrough in the country's industrial OS. The integration of IT application with industrialization begins to have its own "soul".

4. **Relatively backward industrial software**

Industrial software widely used in China, especially software like CAD/CAE used in development and design, sees its core technology controlled by foreign businesses like Dassault, Siemens, Autodesk, and ANSYS. In July 2018, ERI led by the U.S. Department of Defense launched the first group of supporting projects, into which Cadence was the only EDA business to be admitted. Actually, Cadence took the lead in stopping its software sales to ZTE. As the foundation and pillar of advanced industrial products and intelligent manufacturing of China, industrial OS has to be developed on a large scale within the country.

2.4.2   Thriving Automatic Driving

1. **Accelerating automatic driving application in China**

China's automatic driving involves traditional automobile businesses, Internet automobile businesses, Internet businesses, start-up businesses, and college research institutes. They are competing in five segmentation scenarios, namely, ports, industrial parks, open roads, colleges, and highways, with low-speed unmanned distribution vehicles, unmanned sweepers, high-speed self-driving passenger cars, and driverless trucks at ports having been put into use. In correspondence with the L1-L5 standards issued by the American Society of Automotive Engineers, the China Association of Automobile Manufacturers (CAAM) has also put forward five standards, namely, DA, PA, CA, HA, and FA.

2. **Status quo of China's automatic driving technology**

China's automatic driving technology is developing fast, with some application areas in pace with those of the world. It has been applied in environmental perception, precise positioning, decision-making and planning, high-precision mapping, and

V2X (Vehicle to Everything). Baidu has succeeded in applying for over 800 patents in automatic driving technologies, including environmental perception, behavior prediction, planning and control, OS, high-precision mapping, and system security. As for environmental perception, there are laser radar businesses like Shenzhen RoboSense and Hesai Photonics Technology, and webcam businesses represented by AutoX and Beijing Smarter Eye. As for precise positioning, there is Beidou Satellite Navigation system. The precision of Baidu's positioning has reached the level of the centimeter. As for decision-making and planning, there are businesses like Horizon Robotics, Zhejiang Leapmotor, Beijing NavInfo, and Shenzhen SGKS, which are all engaged in R&D of computing chips and platforms. Baidu, Amap, and NavInfo are carrying out the R&D of high-precision mapping. As for V2X technology, Huawei, Alibaba, and Ismartways are carrying out the R&D.

3.   **Automatic driving solution**

After years of technical accumulation, Baidu has established Apollo, a representative platform in automatic driving. In July 2018, at the AI developers' conference, Baidu launched Apollo 3.0, and announced that the 100 L4 automatic buses named Apolong developed jointly by Baidu and King Long were taken off the production line. On September 17, 2018, Wuhan In Driving Tech launched ATHENA automatic driving software system, which integrates functions like high-precision mapping, image recognition, decision control, and route planning, as well as virtual simulation. Beijing TuSimple has developed L4 automatic truck driving, with the webcam as the major sensor, combining laser radar and mm-wave radar, which allows environmental perception, positioning and navigation, decision control. The truck fleet can be put into commercial use in scenarios like port container transfer. Yutong Bus Co., Ltd. announced at its new-energy full-series products launching ceremony in 2018 that some buses it manufactures have the L4 automatic driving capacity for highways and open park commuting roads.

2.4.3   Optimized Intelligent Logistic Resource Allocation

With IT development, China's logistics industry is witnessing integrated development relying on IT. There are trends as follows concerning intelligent logistics.

Focus on standard, open, shared, and platform IT systems. Information of positioning, cargo, and demand at all terminals is analyzed for search of potential value of logistic information, distribution of logistic resources, reduction of storage and transportation capacity, and improvement of distribution efficiency.

Table 2.2 New technology application of intelligent logistics

New technology application	Operation digitalization	Operation automation	Service personalization
Automatic stereoscopic warehouses	✓	✓	–
Perception and recognition technology	✓	✓	–
Automated guided vehicles	✓	✓	–
Portable auxiliary equipment	✓	✓	✓
Drone and unmanned vehicles	–	✓	✓
Intelligent express cabinets	–	–	✓

Establishment of automatic stereoscopic warehouses at all logistic terminals and integration of the latest technologies like things connection perception, Internet, and AI. Automated guided vehicles are used to improve the automation of transportation, sorting, and packing of goods, decrease the error rate of sorting, and improve logistic efficiency.

Intelligent express cabinets and drones and unmanned vehicles adopted for "the last kilometer". That helps to improve the intelligent operation at all terminals and reduce the cost.

New technology application of intelligent logistics is shown in Table 2.2.

Chapter 3
Digital Economy Promoting High-Quality Development

3.1 Overview

Digital economy has become the new drive of China's economic development and a way of constructing the modern economic system. In April 2018, General Secretary Xi Jinping pointed out at the National Cyber Affairs Conference that we should develop digital economy, accelerate digital industrialization, and take information technology innovation as the drive to develop new industries, new business, and new models, so that we can promote new development with new driving forces.

(1) **Digital economy is the new driving force for China's economic growth**. In 2017, the country's digital economic volume reached 27.2 trillion *yuan*, accounting for 32.9% of its GDP, with a year-on-year increase of 2.6%.

(2) **Information consumption is increasing**. By June 2018, the number of China's Internet users amounted to 802 million, including 788 million mobile phone users. The country's information consumption increased from 2.2 trillion *yuan* in 2013 to 4.5 trillion *yuan* in 2017, with an annual increase of 20%, accounting for 10% of total consumption.

(3) **Industrial digitalization is deepening**. The penetration rate of digital economy in service, industry, and agriculture is, respectively, 32.6%, 17.2%, and 6.5%.

3.2 Steady Development of Information Technology

3.2.1 Sound Development of Electronic Information Manufacturing

Digital China Construction Report (2017) issued by Cyberspace Administration of China shows that in 2017, the country's digital economy witnessed its scale at 27.2 trillion *yuan*, with a year-on-year increase of 20.3%, accounting for 32.9% of GDP. Electronic information manufacturing, software and information service, and

© Springer Nature Singapore Pte Ltd. 2020
Chinese Academy of Cyberspace Studies, *China Internet Development Report 2018*,
https://doi.org/10.1007/978-981-15-4043-1_3

communication develop fast. In 2017, the income of information industry amounted to 22.1 trillion *yuan*, with a year-on-year increase of 14.5%.

Electronic information manufacturing is witnessing the increase of both its scale and benefits. *Digital China Construction Report* (2017) shows that in 2017, the country's electronic information manufacturing above designed scale saw a year-on-year income increase of 13.8%. The output of mobile phones, micro-computers, network communication equipment, and color TV sets ranks first in the world. According to data from the Ministry of Industry and Information Technology, in the first half of 2018, electronic information manufacturing above the designed scale saw a year-on-year increase of 12.4, 5.7% faster than that of all industries above the designed scale.

The output of smart devices has increased dramatically. For instance, the output of smart TV sets has amounted to 96.66 million, with an increase of 3.8%; that of industrial robots, 130,000, with an increase of 81.0%; and that of civilian drones, 2.9 million, with an increase of 67%.

Promoted by discrete semiconductor device manufacturing, fixed asset investment in electronic information manufacturing keeps rising again. In the first half of 2018, it witnessed a year-on-year increase of 19.7%, with investment in discrete semiconductor device manufacturing increasing by the highest, up 36.3%, followed by that in integrated circuit (IC) and electronic circuit (EC), whose year-on-year increase was 31.2% and 27.5%, respectively. To build the capacity of the IC industry's sustainable development, the National Integrated Circuit Industry Foundation was established to support the relevant business development. The foundation, as a guide and lever, has promoted the founding of local foundations in Beijing, Shanghai, and Hubei, guiding all kinds of social funds into the IC industry.

The growth rate of the added value of electronic information manufacturing from 2010 to 2018 is shown in Fig. 3.1, and that of fixed asset investment in it in the same period is shown in Fig. 3.2.

Electronic information manufacturing is classified into communication equipment manufacturing, electronic component and electronic exclusive material manufacturing, electronic component manufacturing, and computer manufacturing. Electronic component and electronic exclusive material manufacturing in China has witnessed the highest growth, with its added value growth rate witnessing a year-on-year increase of 15.4%, much higher than the average growth rate of electronic information manufacturing. Then came electronic component manufacturing, with a year-on-year increase of 14.3%. Next come communication equipment manufacturing, and computer manufacturing, with their increase of 13.4% and 7.6%, respectively.

With network users shifting to mobile terminal, the added value of communication equipment manufacturing is much higher than that of computer manufacturing. In the first half of 2018, mobile phone output saw a year-on-year growth of 3.4%; and smartphone output, 4.7%. In the same period, the output of micro-computers saw a year-on-year growth of 0.5%, and that of PCs saw an increase of 1.7%, but that of tablet PCs saw a decrease of 3.2%.

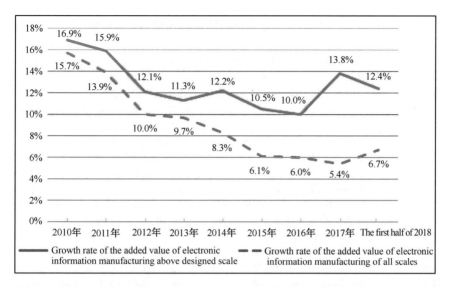

Fig. 3.1 Growth Rate of Added Value of Electronic Information Manufacturing (2010–2018) (This is the data in the first half of the year 2018. The same below) (*Source* Ministry of Industry and Information Technology)

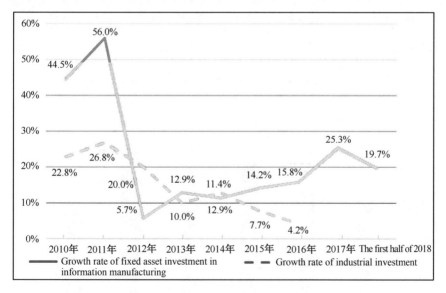

Fig. 3.2 Growth Rate of Fixed Asset Investment in Electronic Information Manufacturing (2010–2018) (*Source* Ministry of Industry and Information Technology)

3.2.2   Steady Growth of Software and IC Service Industry

Data from the Ministry of Industry and Information Technology show that in the first half of 2018, China's software and IT service industry witnessed an income of 2.9118 trillion *yuan*, with a year-on-year increase of 14.4%, and a speed increase of 0.8%. The total profit of the industry amounted to 0.3581 trillion *yuan*, with a year-on-year increase of 10.5%, and a speed decrease of 1.6%.

The income and growth rate of the software industry are shown in Fig. 3.3.

In the whole software industry, the income of information technology service accounts for half, with the highest speed of growth. In the first half of 2018, it amounted to 1.6186 trillion *yuan*, with a year-on-year growth of 17%. The income of software products amounted to 869.1 billion *yuan*, with a year-on-year growth of 13.6%; and that of built-in system software amounted to 424 billion *yuan*, with a year-on-year growth of 6.6%.

Software industry witnesses unbalanced development throughout the country, with that in the east region having the highest income and that in the middle having the highest speed. In the first half of 2018, the software industry in the east saw its income having amounted to 2.3287 trillion *yuan*, with a year-on-year increase of 14.6%; that in the middle saw an income of 122 billion *yuan*, with a year-on-year increase of 19.2%; that in the west saw an income of 3 40.5 billion *yuan*, with a year-on-year increase of 13.5%; and that in the northeast saw an income of 120.6 billion *yuan*, with a year-on-year increase of 8.8%. The income of the software industry in Guangdong, Jiangsu, Beijing, Zhejiang, and Shandong accounted for 65% of the

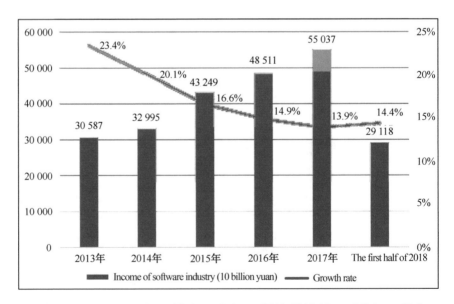

Fig. 3.3 Income and Growth Rate of Software Industry (2013–2018) (*Source* Ministry of Industry and Information Technology)

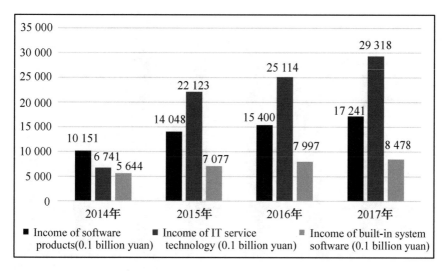

Fig. 3.4 Income of Different Fields of Software Industry in China (2014–2017) (*Source* Ministry of Industry and Information Technology)

national total. In the whole industry, that of the east ranks first, the income of that in Anhui, Hubei, and Henan has the highest growth rate, and that in middle and western regions has to be developed more rapidly.

Software and information technology export is turning better. In the first half of 2018, the export volume was 25.3 billion US dollars, with a year-on-year growth of 2.6%, having increased by 3.3% in growth rate. Outsourcing service export saw a growth of 6.8%, having increased by 2.9% in growth rate; built-in system software export saw a growth of 3.1%. In 2018, the total software export volume accounted for 8.69% of the total, having shifted from negative growth to positive. The Belt & Road Initiative has created a good investment environment for China's software businesses. It is reported that in 2017, the country invested directly 660 million US dollars in information transmission and software and information technology of ten ASEAN countries.

The income of different fields of the software industry in China from 2014 to 2017 is shown in Fig. 3.4 and the export volume and growth rate of that industry from 2013 to 2018 is shown in Fig. 3.5.

3.2.3 Diversified Information Content Services

1. Rising short videos and rapid growth of video payment ratio

According to the 42nd *China Statistical Report on Internet Development*, the number of hot short video APP users has reached 594 million, accounting for 74.1% of the total number of Internet users. In 2017, the market volume of short videos reached

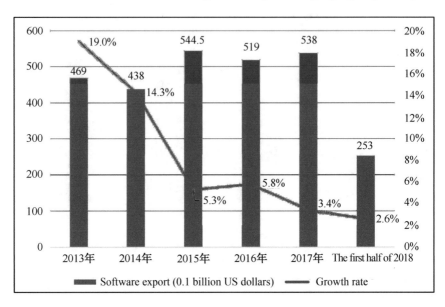

Fig. 3.5 Export Volume and Growth Rate of Software Industry (2013–2018) (*Source* Ministry of Industry and Information Technology)

5.73 billion *yuan*, with a growth rate of 183.9%. The video payment ratio saw rapid growth. Paid videos have become the second income source after advertisement. In 2017, the number of video payers amounted to 130 million, with the payment ratio amounting to 22.5%.

2. **Increasing online game market volume**

In the first quarter of 2018, China's online game market volume was 64.33 billion *yuan*, up 10.3% on a year-on-year basis, including 42.48 billion *yuan* of the mobile online game market volume, with a year-on-year growth of 11.8%; and 21.85 billion *yuan* of computer online game market volume, with a year-on-year increase of 3.1%.[1]

3. **Slowing growth of the number of live streaming users and accelerated live streaming shuffling**

In 2017 and 2018, the growth of the number of live streaming users slowed down dramatically. According to the *42nd China Statistical Report on Internet Development,* by June 2018, the number of live streaming users was 425 million, with the use rate amounting to 53.0%, which saw a decrease of 1.7%. The number increased by 2.94 million in comparison with that at the end of 2017. There is accelerated live streaming shuffling, with the advantages of industrial leaders becoming obvious. In 2017, the number of live streaming businesses decreased by nearly 100, but the total income of that industry increased by 40%, which shows that small live streaming

---

[1] Source: iResearch.

platforms have been knocked out while leaders in that industry can use a large number of users and excellent live streaming content to make a profit. In general, the live streaming industry is developing steadily.

3.2.4 Steadily Growing Telecommunication

Fast-growing mobile data business promotes the rapid growth of telecommunication. In the first half of 2018, that industry saw an income of 672 billion *yuan*, with a year-on-year growth of 4.1%. The total telecommunication saw a revenue of 2.5570 trillion *yuan*, with a year-on-year increase of 132.7%. The growth rate keeps increasing month by month, amounting to 147.4%. The revenue from fixed communication was 195.8 billion *yuan*, with a year-on-year increase of 10%, that from fixed data and Internet was 105.5 billion *yuan*, with a year-on-year increase of 7.2%; that from mobile communication was 476.2 billion *yuan,* with a year-on-year increase of 1.8%; that from mobile data and mobile Internet was 309.5 billion *yuan*, with a year-on-year increase of 12.8%, much higher than the average growth rate of that from mobile communication.

Telecommunication volume and revenue growth rate from 2010 to 2017 is shown in Fig. 3.6.

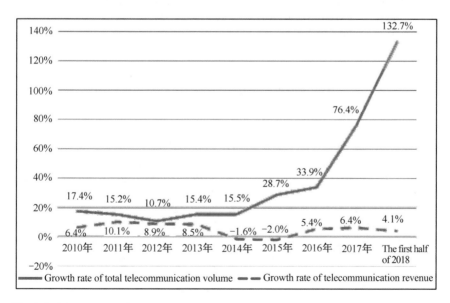

Fig. 3.6 Telecommunication Volume and Revenue Growth Rate (2010–2017)

3.3 Continuous Manufacturing Transformation and Upgrading

3.3.1 Increasing Investment in High-Tech Manufacturing

In August 2018, the National Development and Reform Commission issued *Test on the National Fixed Asset Investment Development Trend by June 2018*. It shows that in the first half of 2018, the investment in manufacturing saw a cumulative growth of 6.8%, with a year-on-year increase of 1.3%; the investment in high-tech manufacturing saw a year-on-year increase of 13.1%, with the growth rate 7.1% higher than that of total investment growth. In the first seven months of 2018, the investment in IC manufacturing saw a year-on-year increase of 67.9%; that in medical diagnosis, monitoring and treatment equipment saw a year-on-year increase of 65.1; and that in optoelectronic devices, 45.5%. The output of new-energy cars, industrial robots, and ICs saw a growth of 68.6%, 21.0%, and 14.5%, respectively. Emerging products like service robots, smart TV sets, 3D printing equipment, bio-based chemical fibers, and solar cells maintain a high growth rate. *Intelligent Manufacturing Market Prospect and Investment Strategic Planning Analysis Report* shows that in the first half of 2018, high-tech manufacturing actually attracts 43.37 billion *yuan* of foreign investment, with a year-on-year increase of 25.3%. Foreign investment in electronic and communication equipment manufacturing, computer and office equipment manufacturing, and medical instrument and meter manufacturing, respectively, saw a growth of 36%, 31.7%, and 179.6%.

3.3.2 Deepening Integration Between Informatization and Industrialization

In June 2017, at the conference of the launching of the national standards for the integrative management of informatization and industrialization, GB/T 23000–2017, namely, *Foundation and Terminology of Integrative Management of Informatization and Industrialization* and GB/T 23001–2017, namely, *Requirements for Integrative Management of Informatization and Industrialization*, were officially issued. Throughout the country, 4,300 businesses have adopted the standards, with their operation cost having decreased by 8.8%, and their profit having increased by 6.9%. Businesses are voluntarily adopting the standards, with increasing efficiency.

According to data from Informatization and Industrialization Integration Service Platform, in 2017, China's index of integration between informatization and industrialization reached 51.8%. 114,919 businesses have participated in the assessment of the integration. Of the assessment, single item coverage accounts for 47.7%, initial construction accounts for 33%, integration upgrading accounts for 15.2%, and innovation breakthrough accounts for 4.1%, as is shown in Fig. 3.7. In the first quarter of

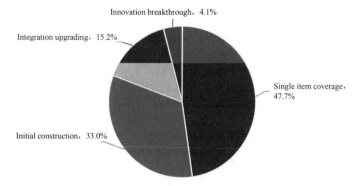

Fig. 3.7 Stages of China's Integrative Development of Informatization and Industrialization (*Source* Informatization and Industrialization Integration Service Platform

2018, the popularization rate of entrepreneurship and innovation platforms reached 71.5%; that of digitalized R&D design tools, 67.4%; that of industrial e-commerce, 58.7%; and that of key process control, 47.8%. 33.7% of businesses have adopted network coordination, but the proportion of service, smart and personalized manufacturing businesses remains low, 24.4%, 6.7%, and 7.3%, respectively. China's key indicators of integration between informatization and industrialization are shown in Table 3.1.

3.4 Digital Technology Facilitating Rejuvenation of Rural Areas

3.4.1 Improving Information Service for Farmers, Agriculture and Rural Areas

The coverage of the "information entering households project" is expanding, with 18 provinces (and municipalities directly under the Central Government) involved. Farmer-friendly information stations are being founded to provide services covering charity, convenience service, e-commerce, and training experience. They have been founded in 204,000 villages, which account for one-third of the national total. It is expected that such information stations will cover all villages of the country in the coming 2–3 years.

Information resource sharing and openness are deepening. After its founding, the Ministry of Agriculture and Rural Affairs set up four agricultural and rural area information-sharing platforms, covering agricultural product quality and safety tracing, agricultural veterinary medicine, key agricultural product market information, and new-type agricultural entity information reporting. Big data pilot programs have been launched for 8 agricultural products in 21 provinces (and municipalities

Table 3.1 China's key indicators of integration between informatization and industrialization

Quarter	Digitalized R&D design tool popularization rate (%)	Critical process control rate (%)	Industrial e-commerce popularization rate (%)	Entrepreneurship and innovation platform popularization rate (%)	Smart manufacturing readiness rate (%)	Percent of businesses having adopted network coordination (%)	Proportion of service manufacturing businesses (%)	Proportion of personalized manufacturing businesses (%)
2nd quarter of 2017	63.2	46.4	55.1	60.0	5.6	31.0	24.3	7.3
3rd quarter of 2017	63.3	46.4	55.4	70.4	5.6	31.2	24.3	7.3
4th quarter of 2017	66.4	47.4	58.0	70.7	6.6	33.7	24.3	7.3
1st quarter of 2018	67.4	47.8	58.7	71.5	6.7	33.7	24.3	7.3

directly under the Central Government) to improve the monitoring and alarming system; the wholesale price index of agricultural products is published every day, the supply and demand report for 19 agricultural products and supply and demand balance sheet are launched every month so that data management service can guide production and sales.

Farmers are trained in using APPs for them and for agriculture and rural areas. In the past 3 years, the Ministry of Agriculture and Rural Affairs has carried out training for farmers so that they can use their mobile phones for knowledge of market demand and agricultural technology. Mobile phones are their new agricultural tools.

3.4.2 Digital Technology Upgrading Agricultural Production Modes

"Internet + agriculture" shapes new agricultural modes. (1) Cross-border agriculture. Agriculture is being integrated with tourism, finance, and culture. It is no longer a single operation mode like farming or animal husbandry. (2) Custom-tailored agriculture. Platforms are constructed between farmers and consumers, so that through real-time communication, farmers can produce what consumers demand, and consumers can learn about the production process, and hence both parties are well-informed of each other and their trust in each other is enhanced. (3) Safe agriculture. Systems of agricultural product quality tracing are established to enable them to have their ID. Each product has its QR code, and by scanning this code consumers can know the production, processing, storage, transportation, and sales of the products.

3.4.3 Rapid Development of E-Commerce in Rural Areas

On August 20, 2018, the Central Committee of the CPC and the State Council issued the *Guiding Opinions on Three-year Action of Winning the Battle against Poverty*, according to which actions of poverty alleviation through communication and express's entering rural areas will be accelerated, logistic distribution systems will be improved, and a safe and convenient communication and transportation network will be set up with internal and external, and village and township connection, and with buses to villages. According to statistics published by the Ministry of Commerce, in 2017, the online retail volume in rural areas of China reached 1,244.88 billion *yuan*, up 39.1%. In 832 national-level poverty-stricken counties, it reached 120.79 billion *yuan*, up 52.1%. It is expected to amount to 1.6 trillion *yuan* in 2018. By the end of 2017, there were 9,856,000 online stores in rural areas, 1,693,000 more than in 2016, up 20.7%, providing jobs for 28 million people. Among these stores, the online retail volume of physical products from rural areas was 7 82.66 billion *yuan*, up 35.1%, accounting for 62.9% of the total online retail volume in rural areas.

3.5 Digital Innovations in Service Industry

3.5.1 Life Service Digitalization Improving Life Quality

E-commerce is an engine of modern service industry development, with its trans-action volume growing fast. Statistics from the Ministry of Commerce show that in 2017, the e-commerce transaction volume of China amounted to 29.16 trillion *yuan*, up 11.7%. Statistics from the National Bureau of Statistics show that from January to June 2018, the online retail volume of the country amounted to 4,081 billion *yuan*, up 30.1%. That of physical commodities amounted to 3,127.7 billion *yuan*, up 29.8%, accounting for 17.4% of the total retail volume of consumer products. In the online retail volume of physical products, those concerned with eating, wearing, and daily use account for 42.3%, 24.1%, and 30.7%, respectively. According to the latest data from the Ministry of Industry and Information Technology, from January to July 2018, the income from e-commerce platforms amounted to 177.6 billion *yuan*, up 37.8%, witnessing a rapid growth. By the end of February 2018, the number of e-commerce APPs amounted to 419,000, nearly 30,000 more than that of the end of 2017. The number of downloads of the third-party market APPs was 56.1 billion. E-commerce APPs are active, with Taobao, JD, and TMALL being the most active. On the other hand, the activity of e-commerce sees uneven distribution, traditional e-commerce platforms having the highest traffic. There have emerged a number of socializing e-commerce businesses, with 223 million users of socializing retailing, contributing to the expansion of e-commerce.

China's total e-commerce volume and growth rate from 2011 to 2017 are shown in Fig. 3.8.

The number of "Internet + tourism" users is increasing, and personal demand is the new highlight of tourism. Online tourist platforms began to provide more services for stock users, and China's tourists have shifted their demand from sight-seeing to relaxation and local experience, so they have a higher demand for tourist products. Online tourist platforms have shifted their operation from online agency to product development, and from competition over traffic to competition over service. Family tours and tailored FIT and other products are more focused on consumers' personal demand.

"Internet + education" promotes the balance of quality education resources. According to a report by Prospective Industry Research Institute, in 2017, China's online education market volume reached 278.81 billion *yuan*, up 26%. It is expected to surpass 300 billion *yuan* in 2018. According to Baidu big data of education, nearly 60% of Internet learners are living in third-tier or fourth-tier cities and rural areas, and users in remote areas long for quality Internet learning resources more than those in other areas. Online education goes beyond time and space limit to traditional in-class education, so that learners in poverty-stricken areas can access the educational resources in the first-tier cities. Thus, equality in education has been facilitated. Technologies of speech recognition, online-assessment, and live streaming and

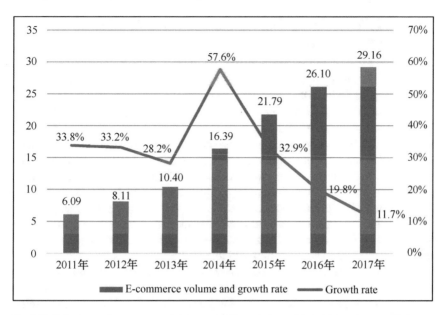

Fig. 3.8 China's Total E-commerce Volume and Growth Rate (2011–2017) (*Source* Ministry of Commerce)

interaction have been upgraded, forming a closed loop and satisfying the demand for teacher-learner interaction and Q&A.

"Internet + medical care" is witnessing the period of new strategy opportunity. On April 25, 2018, the General Office of the State Council issued Opinions on Promoting the Development of "Internet plus Health Care (guobanfa 2018 No. 26), in which the service system of "Internet + Health Care" is proposed. It also proposes the improvement of the supporting system for the service system, and the reinforcement of its supervision and security guarantee. With the increase of their income, people have an increasing demand for medical care and health. "Internet + medical care" can improve medical care efficiency, meeting people's increasing demand for it. There are three types of "Internet + medical care", namely, service including online diagnosis and registration, information (mainly inquiry about information and consulting), and transaction (online medicine shopping). In 2017, the online medical care market volume reached 22.3 billion *yuan*, and the number of its users reached 253 million, accounting for 32.77% of the total Internet users.

3.5.2 *Digitalization of Production Service as the New Driving Force of Real Economy*

"Internet + logistics" is the breakthrough point for cost reduction and efficiency improvement of China's logistics. In August 2017, the General Office of the State Council issued *Opinions on Facilitating Logistic Cost Reduction and Efficiency Improvement to Promote Real Economy Development*, which requires to develop new industries and modes such as "Internet+" vehicle-goods match, "Internet+" transportation capacity optimization, "Internet+" transportation coordination, and "Internet+" warehousing trading with advanced information technologies like the Internet, big data, and cloud computing. Policy support will be offered to cultivate a number of leading businesses and boost NTOCC pilot projects. Internet platforms will be built to innovate logistic resource distribution modes, expand resource distribution, facilitate real-time sharing and intelligent match of freight transportation, and reduce detour, empty transportation, and idle logistic resources.

"Internet + logistics" is taking on many modes. First, online platforms facilitate the efficient match between freight wagons and freight transportation resources, help wagons to find freight and freight to find wagons to carry to reduce the empty return ratio of freight wagons. For instance, Yumanman Platform, based on the National Trunk Logistic Scheduling System, succeeded in intelligent wagon-cargo match, real-time scheduling, standard pricing, and map tracing. The number of heavy truck drivers registered till now has reached over 3.9 million, and that of the owners of heavy trucks, 0.85 million. There are 240,000 daily transaction waybills and 1.5 billion *yuan* of daily matching transaction volume. Secondly, the crowdsourcing logistic mode is witnessing the improvement of distribution efficiency. One of the crowdsourcing platforms providing express service within one city was rrkd.cn, on which there are millions of freelancers, who provide delivery, purchasing, and assistance for nearly thousands of users in nearly 70 cities. Within one city, an assigned man is expected to deliver directly, and others can deliver by the way. Compared with traditional modes, this mode has helped to reduce the time of collection and delivery. Thirdly, logistic facilities are shared, for instance, delivery end stores' facilities within one city. In some logistic park areas, through online platforms, the owner of goods, logistic businesses, supporting service businesses, and individual workers are connected with one another, forming new logistic biospheres.

Crowdsourcing platforms oriented to manufacturers are developing fast, changing the traditional production modes. In recent years, crowdsourcing platforms have been widely used in industries. Freedom on them and their openness and integration have changed not only the production mode but also people's working and living styles, reallocated resources, and improved their use efficiency. Crowdsourcing platforms can be classified into manufacturing technology, software, R&D, and sales crowdsourcing. IT service, comprehensive task, and part-time free mission crowdsourcing platforms are typical crowdsourcing platforms in China.

3.5.3 Cross-Border E-Commerce Facilitating New Foreign Trade in the Digital Era

Cross-border e-commerce is a strategy for the internalization of China's e-commerce retail. The rapid development of digital economy is contributing to the country's integration into the global economic development in a bigger scope. By the end of 2017, the number of Haitao users had amounted to 65 million, and it is expected to rise to 88 million in 2018. In 13 comprehensive pilot zones of cross-border e-commerce in Hangzhou, the administration system with 6 subsystems and 2 platforms as the core has been set up, their experience in 12 aspects copied and promoted throughout the country. According to statistics, in 2017, the import and export volume of 13 zones amounted to over 330 billion *yuan*, a doubled figure of that of the previous year, far higher than the average growth of foreign trade of the country. Cross-border e-commerce is a new highlight in foreign trade growth. Shenzhen, Guangzhou, and Hangzhou are the top three cities in its transaction volume, which is, respectively, 49.166 billion US dollars, 22.77 billion US dollars, and 9.936 billion US dollars, with the growth rate being 21.84, 55.1, and 22.5%. On July 13, 2018, the State Council established another 22 comprehensive pilot zones of cross-border e-commerce, most of them in the middle-west and northeast of China. This brings new opportunities for the development of the middle-west and northeast regions. There are 35 pilot zones of cross-border e-commerce.

The domestic market is open to the world, and the total import of cross-border e-commerce is more than the total export volume. According to *Report on China's E-commerce Development* (2017), the cross-border e-commerce import and export volume was 90.24 billion *yuan*, with a year-on-year growth of 80.6%. In it, the import volume was 56.59 billion *yuan*, with a year-on-year growth of 120%; the export volume was 33.65 billion, with a year-on-year growth of 41.3%. It is obvious that the volume and growth rate of cross-border e-commerce imports are higher than those of cross-border e-commerce export.

3.6 Financial Technology Facilitating New Industrial Reforms

In the digital era, with the development and application of cloud computing, big data, AI, and blockchain, financial technology develops fast, and it is shaping the ecosystem of the financial industry. Mobile payment is penetrating all scenarios, Internet financial product innovations keep emerging, investment/funding is more centralized, convenient, and customized, and more and more users will enjoy equal and safe financial services. Meanwhile, industrial supervision is faced with new challenges.

3.6.1 Mobile Payment Being Done in All Scenario Consumption

According to the *42nd China Statistical Report on Internet Development*, by 2017, the mobile payment reached 202.93 trillion *yuan*, ranking first in the world. The number of mobile payment users was 527 million, 57.83 million more than that in 2016, with an annual growth rate of 12.4%. The mobile payment use rate was as high as 70%. The mobile payment accounts for 82.4% of the total payment. Alipay is promoting the "going out" of technology, and its payment service is localized in India, Thailand, South Korea, the Philippines, Indonesia, Malaysia, Pakistan, and Bangladesh. Such payment has served over 870 million people.

Mobile payment covers almost all scenarios. As for specific consumption scenarios, living consumption is the area with the highest mobile payment popularization rate, as high as 97.1%. Then comes ticket payment, accounting for 80.4%. That in tourism is 64.7%. Next is public payment and amusement download, with the mobile payment rate of over 50%. According to *China's Digital Economy: a leading global power* published by McKinsey Global Institute, China is a major force in global mobile payment, with the transaction volume 11 times that of the United States.

3.6.2 Standardization of Online Crowdsourcing and P2P Loaning

In recent 2 years, with the increasing number of Internet financial platforms with problems, a series of supervision policies have been launched and implemented. In December 2017, the Office of P2P Loaning Risk Rectification Leading Group issued the *Notice of Acceptance of P2P Loaning Risk Rectification*, which requires that the local authorities should complete the registration of P2P businesses within their jurisdiction by the end of April of 2018, and all P2P businesses by the end of June of the same year. Specific rectification and registration schedule was issued. In December 2017, the Office of Internet Financial Risk Rectification Leading Group and the Office of P2P Loaning Risk Rectification Leading Group issued the *Notice of Standardization and Rectification of Cash Loaning*, which clarifies the principles for cash loaning, the way of online loaning platforms' involving in cash loaning and illegal practice in that area.

3.6.3 Emerging Internet Financial Products

A variety of products have emerged with the deepening of Internet finance. For instance, the combination of Internet finance and artistic product market has led to the emergence of art product crowdsourcing, mortgage loaning, foundations, and options. The combination of Internet finance and sharing economy has led to the emergence of sharing economy insurance, which connects the supply chain finance of the supplier, manufacturer, distributor, retailer, and ultimate user. These products have, to some extent, promoted the development of financial Internet and its related industries, but on the other hand, the financial risk cannot be ignored. At present, financial technology is oriented to possible financial risks, and new technologies like big data, blockchain, intelligent risk management, and biometric identification are used to solve such risks, which has seen positive effects.

Column 3: Technology Stack Independently Developed by Ant Financial

Technology stack independently developed by Ant Financial is the technical foundation of financial transaction safety and risk management. China no longer relies on basic foreign software in that respect. Ant Financial has established the complete technology stack covering resource management of basic facilities like servers, networks and memory, system platforms of distributed databases and middleware, big data and mobile development frameworks, and new technical areas like blockchain, risk management, and biometric identification, ensuring the safe transaction by 87 billion Alipay users across the world. It witnessed the unprecedented 256,000 peak transactions per second on November 11, much higher than the figure of its kind published by other leading payment platforms of the world.

The risk management technology independently developed by Ant Financial has developed into risk brain technology. In comparison with traditional risk management, Ant Financial Risk Brain has upgraded the traditional "post-incident" risk discovery to the combination of "in-the-incident" and "pre-incident" risk alarming, so that financial institutions can take measures to reduce the loss to the minimum. Today, it can scan nearly 460 million transactions every day, with the loss rate at only five per 10 million. Its risk management covers account safety, network fraud, marketing cheating, and loaning cheating so that tens of millions of users' payment safety is guaranteed. Ant Financial Risk Brain has been exported as copyright for the convenience of more than ten e-wallets and e-commercial platforms to ensure the fund and account safety of users across the world.

The intelligent supervision technology system of Ant Financial Risk Brain can monitor in real-time five million businesses of China that are suspected of doing financial business, with the rate of identification of risks like illegal fund-raising and pyramid sales as high as over 93%, providing risk identification and tackling services for over 20 local financial supervision departments, and localization customization for five clients. In collaboration with public security authorities, Ant Financial has combated 1,000 black gangs, recovering nearly 10 million *yuan* of fund for users in one year. It cooperates with nearly 600 banks and other financial institutions, and over 1,000 Internet companies in providing risk management solutions for partners and micro and small businesses, monitoring and guaranteeing nearly one billion third-party transactions every day.

Looking into the future, we find that the development of Internet finance will bring financial service closer to people's life. On the other hand, financial innovation poses new challenges to and has requirements for supervision innovation. There is an immediate need to make use of IT to improve the supervision capability, so that supervision can cover all scenarios, chains, and processes. In this way, financial technology can better prevent national financial risks, guarantee national financial security, and promote financial development.

Chapter 4
Steady Opening-Up of Governmental Data

4.1 Overview

Since the 1980s, China has been attaching importance to and promoting the fast development of e-government building, having launched a host of e-government information projects like Finance Project and Government's Access to the Internet. The country's e-government network framework, e-government systems and regulations have been set up. Thanks to e-government informatization, governmental sectors, governmental agencies, and relevant public organizations have collected and stored a large amount of information, and public information publicity is a concern of both the government and society, having witnessed four stages, i.e., government's access to the Internet, information publicity, information sharing, and public information publicity.

(1) **Effective governmental information publicity**. The government publishes its accountabilities list, authorities list, information publicity catalog and public service list on the information publicity website. China has complete polices, institutions, and measures for governmental information publicity. The provincial-level government has seen the effective opening of catalog list, and the prefectural-level government is also seeing some effect in that respect.

(2) **Information-sharing driven by governmental service being improved and multiple-source information integration to be improved**. During the governmental information sharing, departments of different levels can share their information, optimize the governmental resource allocation, and improve work efficiency. With better and better policies, institutions, and measures, governmental information sharing has seen great effect but deep information integration has to be enhanced.

(3) **Great momentum of public information publicity, which sees initial effects**. During the governmental information publicity, public information sharing and integration is promoted to manifest the economic and social effect of public information. China has launched its public information publicity, contributing

© Springer Nature Singapore Pte Ltd. 2020
Chinese Academy of Cyberspace Studies, *China Internet Development Report 2018*,
https://doi.org/10.1007/978-981-15-4043-1_4

to the improvement of governmental capacity and public service and facilitating industrial development.

4.2 History of Public Information Publicity

4.2.1 Government's Connection to the Internet

The government's connection to the Internet refers to the government's functions' performance on the Internet. China's government's connection to the Internet began at the end of the 1990s. In April 1998, Qingdao Municipality opened its website www.qingdao.gov.cn, which was China's first governmental website in its strict sense. On it, information could be released and collected and some office work could be done. In January 1999, China Telecom cooperated with over 40 information authorities like the Economic Center of Ministry of Foreign Trade and Economic Cooperation in launching the Government's Connection to the Internet and opening the major website www.gov.cninfo.net and portal website www.gov.cn. Today, there are 22,206 governmental websites (including www.gov.cn). Among them, there are 1839 websites of the State Council and its internal and vertical administrative institutions, 32 provincial governmental portal websites, 2265 provincial departmental websites, 518 municipal portal websites, 13,614 municipal departmental websites, 2754 county-level portal websites, and 1183 websites below the county level.

4.2.2 Governmental Information Publicity

With the government's connection to the Internet, governmental information is being publicized. In March 2004, the State Council issued *Outlines of Promoting the Administration by Law*, according to which administrative decision and management and governmental information are taken as the important content of promoting the administration by law. In May 2007, the State Council formulated *Rules of Governmental Information Publicity*, which stipulates the governmental sectors' accountabilities, and the range and procedure of information publicity to make governmental information institutionalized and standardized. By December 2017, 31 provinces (and municipalities and autonomous regions) and Xinjiang Production and Construction Corps have published their provincial-level governmental sectors' power list and accountability list.

4.2.3   Governmental Information Sharing

With China's government's connection to the Internet and governmental information publicity, governmental information sharing is put on the agenda. In 2003, eight departments including the Cyber Affairs Office of the State Council, General Administration of Customs, and State Taxation Administration decided to launch pilot projects in basic information exchange of import and export businesses. All competent authorities from the central government to local governments have proposed the demand for governmental information sharing in different areas. In April 2013, the National Development and Reform Commission, in collaboration with the other six ministries, issued *Guidelines on Further Enhancing Governmental Departments' Information Sharing Construction Management,* which marks the official beginning of China's governmental information sharing. In September 2019, the State Council issued *Interim Measures for the Administration of Sharing of Government Information Resources,* which defines the scope and responsibilities of information sharing, and the regulations and requirements for information sharing administration, coordination, assessment, and supervision. In May 2017, General Office of the State Council issued the *Implementation Plan for Governmental Information System Integration and Sharing,* which defines the key tasks and implementation paths for accelerating governmental information system integration and sharing and facilitating the interconnection of information systems between the State Council departments and local governments. In May 2018, Premier Le Keqiang presided over the executive conference, at which he pointed out that by 2019, at least 90 and 70% of governmental services at the provincial and municipal level and county level, respectively, should be done on the Internet, which will mark a new chapter and breakthrough of China's governmental information sharing.

Since the abovementioned policies were implemented, local governmental departments have done much work by making breakthroughs, having created conditions for governmental information sharing. For instance, some ministries have built and now share four basic national databases, namely, Basic Population Information Database, Basic Corporation Information Database, Basic Natural Resource and Space Geography Information Database, and Macroeconomic Information Database, which provide a guarantee for governmental information sharing and cross-sector coordination. Basic Population Information Database was built jointly by Ministry of Education, Ministry of Civil Affairs, Ministry of Human Resources and Social Security, and former National Health and Family Planning Commission, with Ministry of Public Security as the coordinator. It contains 1399 million effective information entries of 13 data items, seven of which enjoy 100% of the collection rate. It makes possible coordinated management of China's basic information, solving the problem in reality and validity of information concerning births and deaths collected by different departments before. National Development and Reform Commission, through the four platforms, namely, credit information sharing, 12,358 price supervision, online approval and supervision of investment projects, and public resource transaction, has succeeded in streamlining administration and delegating power, combining management with delegation of power, and optimizing service, which has stimulated the vitality of market and creativity of society.

4.2.4 Public Information Publicity

Public information publicity is a concern of both the government and the public. In August 2013, the State Council issued *Opinions on Promoting Information Consumption and Expanding Domestic Demand*, in which the concept of "public information" was put forward. The document proposes the blueprint for the formulation of regulations on public information publicity and sharing and the acceleration of construction of pilot provinces and municipalities in that respect, which marks the beginning of public information publicity. In August 2015, the State Council issued the *Action Outline for Promoting the Development of Big Data* to promote the governmental information system and public data interconnection, publicity, and sharing. In February 2017, at the thirty-second conference of Central Leading Group for Comprehensively Deepening Reforms, *Opinions on Promoting Public Information Publicity* was passed, contributing to public information publicity. In January 2018, Office of the Central Leading Group for Cyberspace Affairs, National Development and Reform Commission, and Ministry of Industry and Information Technology jointly issued *Work Plan for the Pilot Program of Opening of Public Information*, in which the task of public information publicity is defined. Beijing, Shanghai, Zhejiang, and Guizhou have publicized some data on their provincial-level public information publicity platforms, having formed some effective regulations and experiences and played a leading and demonstrating role in that respect.

In general, China's public information publicity is witnessing steady progress. First, more departments are involved and all publicity bodies contribute to the cooperation. For instance, there are 53 governmental departments in Beijing and 43 in Shanghai involved in public information publicity. Secondly, there are more types of information published. The opened public information resources now cover road transportation, schooling, and health and medical care, and they can be categorized into primary data, processed data, data service API, and applications that can be used directly. Thirdly, unified local publicity platforms are running smoothly and their functions can be fulfilled in a central way. For instance, in January 2017, Guiyang Municipal Government's data publicity platform was launched, with 634 data sets and 101 APIs opened for free, involving over 50 municipal departments and public institutions. Fourthly, there are obvious effects in public information publicity, with breakthroughs in governmental administration, industrial development, and public service.

4.3   Remarkable Effect of Governmental Information Publicity

4.3.1   Better and Better Governmental Information Publicity Policies

In March 2007, the State Council issued *Regulation of the People's Republic of China on the Disclosure of Government Information*, requiring that administrative authorities should publicize their information, the people's government of different levels should establish and perfect the governmental information publicity evaluation system, social reviewing system and accountability system, so that regular evaluation and reviewing can be done. In January 2010, the State Council issued *Opinions on Governmental Information Publicity upon Application*, which requires that the application method should be adjusted on the principle of "one application for one matter", the active publicity of governmental information should be enhanced and the publicity should be improved. In December 2015, the General Office of the State Council issued *Notice of the State Council's Power and Accountability List Compilation Pilot Program*, according to which power and accountability list compilation and publicity should be evaluated in different regions. It covers whether standard accountability and power list are publicized as required, and whether matters, basis setting, accountability, and power types are publicized. In February 2016, General Office of the CPC Central Committee and General Office of the State Council issued *Opinions on Comprehensively Advancing the Work of Open Government*, which requires that governmental service centers should cooperate with online service halls to promote governmental service's expansion onto the Internet and the publicity of the power list, accountability list, and negative list and to establish and improve the dynamic list adjustment publicity mechanism.

At the local level, in May 2016, the Beijing Municipal Government issued the *Work Focuses of Beijing Governmental Information Publicity* in 2016, which stressed that new breakthroughs in information publicity, policy interpretation, response to the public concern, and public involvement should be made to facilitate reform deepening, economic development, life improvement, and governmental construction. In May 2018, Zhejiang Government issued *Work Focuses of Zhejiang Governmental Information Publicity in 2018*, which requires enhancing active publicity and response interpretation to improve the efficiency of governmental service and deepen "Internet + Governmental service". Facilitated by notification of random inspection, all regions and departments have enhanced their organization and leadership, clarified work division, and perfected institutions and regulations. Thus, the operation and management of governmental websites throughout the country have been improved.

4.3.2 Remarkable Provincial Governmental Information Publicity Achievement

Thirty-two provincial governmental websites (including that of Xinjiang Production and Construction Corps) have provided their power and accountability lists as required by the State Council. In particular, 13 websites, including those of Beijing and Shanxi, have provided complete power and accountability lists, information publicity catalogs and public services, with every list having a clear and complete classification guide on it. Such kind of excellent catalog coverage capacity building requires the governments to provide catalog list portals, which are accessible to the public, who can know about the governmental responsibility boundary and about services that the government can provide.

4.3.3 Good Prefecture-Level Governmental Information Publicity

According to *Report on China's Local Governments' Internet Service Capacity Development* issued in 2018, among the country's 334 prefecture-level municipal government websites, Changchun and Baishan of Jilin Province and Dalian of Liaoning Province have high catalog coverage capacity, and 206 prefectures have their catalog coverage capacity higher than the national average, accounting for 61.68%. Generally speaking, the catalog coverage of prefecture-level municipal governmental websites has to be improved. In the four dimensions of the coverage capacity, governmental information publicity catalog scores 91.07%, which is the highest; power and accountability list scores 66.71 and 64.11%, respectively; public service list scores 31.12%, which is the lowest. The construction of prefecture-level municipal government websites' catalog list portals is supposed to be the same as that of the provincial-level ones to ensure that the public can find the expected data in a convenient and fast way.

4.4 Governmental Information Sharing Being Improved

4.4.1 Governmental Information Sharing Policies

In September 2016, the State Council issued *Interim Measures for the Administration of Sharing of Government Information Resources*, which proposes explicit requirements and systematic methods for China's information sharing. On September 25 that year, the State Council issued *Guiding Opinions on Accelerating "Internet + Governmental Service"*, which stresses the acceleration of mutual acknowledgement of governmental information, the breaking of data barricade, and the interconnection and sharing of data and information between departments and between different levels. In May 2017, General Office of the State Council issued *Notice of Issuing the Implementation Plan for Governmental Information System Integration and Sharing*, which stresses the acceleration of governmental information system integration and sharing. In June 2018, the office issued *Notice of Deepening the Reform Implementation Plan for One Website, One Gate and One Time of "Internet + Governmental Service"*, which stresses cross-level, cross-regional, cross-systematic, cross-departmental and cross-business interconnection, coordination, and sharing. In July 2018, the State Council issued *Guiding Opinions on Accelerating the Construction of the National Integrated Online Governmental Service Platform* to give full play to the role of the national governmental service platform as a public portal, channel, and support. With data sharing as the core, China will enhance cross-regional, cross-systematic, cross-departmental, and cross-business coordination, promote governmental service publicity and data publicity and sharing oriented to the market and individuals, and to deepen "online link", "data link", and "business link".

4.4.2 Great Achievements of Governmental Information Sharing

1. **All-in-one-network governmental service deployment**

In China, 31 provincial governmental service platforms have been set up, and 30 State Council sectors have built and opened their governmental service platforms. Twenty regions have set up online governmental service systems above the provincial, municipal, and county levels. Zhejiang, Guangdong, and Guizhou have set up five levels of online governmental service systems, namely, provincial, municipal, county, township, and village levels. Among 22, 152 provincial issues with administrative permission, 16, 168 can be submitted online for pre-examination, accounting for 72.98%, with their process time shortened by 24.96%. Some regions and sectors, relying on the innovative governmental service on platforms, have launched and improved reforms and measures like one-time entry solution and online examination. Zhejiang launched the reform "once at the most" in 2016, and promoted "one-window and integrated service" so that individuals and businesses only need to come to one window for service and get it at once. Figure 4.1 shows the APP page of the Zhejiang mobile service. In 2018, Shanghai promoted "all-in-one-network" online and offline governmental service oriented for businesses and individuals, who can enjoy the service in one network at once. Figure 4.2 shows the home page of Shanghai's "all-in-one-network" service.

Fig. 4.1   APP page of Zhejiang mobile service

Fig. 4.2   Home page of Shanghai's "All-in-One-Network" service

2. **Improved law enforcement effects**

Administrative regulation information is widely used in safeguarding public security, punishing tax evasion, penalizing law breach by vehicles, and forbidding unlicensed merchants. For instance, the National Gold Shield Project, through the construction of three-level information networks and the extension of terminals, ensures information sharing among the Ministry of Public Security, provincial and prefecture-level municipal public security authorities, and grass-root public security stations, playing an important role in criminal pursuit, trafficking combat, and motor vehicle robbery combat.

3. **Convenient and fast service for the public**

By the end of March 2018, the number of national social security cardholders had amounted to 1.12 billion, the popularization rate being 80.6%. The cardholders are from all regions. Thanks to central, provincial, and municipal networks, human resource and social security information platforms with a wide coverage and high application level have been built throughout the country. According to the latest data released by National Healthcare Security Administration, by the end of 2018, there were 10,458 designated medical institutions across provinces, with 808,000 payments (including 102,000 payments from the New Rural Co-operative Medical System) settled across provinces, covering medical expenses of 19.46 billion *yuan* (including 1.76 billion *yuan* paid from the New Rural Co-operative Medical System), and fund payments of 11.42 billion *yuan* (including 750 million *yuan* paid from the New Rural Co-operative Medical System), accounting for 58.7%.

Figure 4.3 shows the national New Rural Co-operative Medical System information platform.

4.4.3 *Information Integration to Be Improved*

Public information integration has begun. Decision-making departments from the government will fulfill the decision, administration, and service responsibilities better through multiple-source, cross-area, and cross-sector information integration. For instance, in February 2018, Xianyang Municipality of Shaanxi Province launched *Xianyang's Implementation Plan for Big Data Development (2018–2020)*, which proposes that the intelligent Xianyang geographical information system and cross-sector information sharing and monitoring gridding and elaborative gridding forecast shall be relied on together with cloud maps, radars, automatic stations, and regional stations to ensure horizontal integration of disaster-prevention and reduction data, to set up the city's big data bank and information system of disaster prevention and reduction, and to foster the commanding platform for natural disaster monitoring and early warning.

Fig. 4.3 New rural co-operative medical system information platform

4.5   Strong Momentum of Public Information Publicity

4.5.1   *Public Information Publicity Route Being Clarified*

In August 2015, the State Council issued the Action Program on Promoting Big Data Development, which requires that by the end of 2018, China will have established a unified governmental data publicity platform. In February 2017, Opinions on Facilitating Public Information Publicity was put into implementation. It proposes that comprehensive deployment and pilot projects should be combined and the reform in that respect should be facilitated in accordance with laws. Later, Work Program for Public Information Publicity Pilot Projects specified the requirements for public information publicity, defining Beijing, Shanghai, Zhejiang, Fujian and Guizhou as the pilot regions for public information publicity. The regions have been expected to make specific plans of implementation covering the establishment of relevant platforms, clarification of publicity scope, improvement of data quality, data use, regulation making and security guarantee, so that they can provide experience for other regions of the country. They have been asked to finish all work concerned before the end of 2018.

4.5.2   Rapid Public Information Publicity

By the first half of 2018, China had launched 46 provincial and prefecture-level platforms with basic characteristics of governmental data publicity, including 15 provincial ones and 31 prefecture-level ones. Information publicity progress in Beijing, Shanghai, Zhejiang, Guizhou, and Shenzhen will be taken as an example in the following part.

1.   **Establishment of unified platforms for data publicity**

By October 2018, platforms for data publicity had been set up in all provinces (and autonomous regions and municipalities directly under the Central Government) of China. Those in Beijing, Shanghai, Zhejiang, Guizhou, and Shenzhen are shown in Table 4.1. Data publicity platforms show data set publicity, dynamic data display, convenient searching, interaction and communication, and tool service.

(1)   Publicity of data sets on platforms

The platforms in Beijing, Shanghai, Guizhou, and Shenzhen are unified and exclusive. That is, publicized data are collected onto one special platform for publicity on the home page. In Zhejiang, the platform is a unified built-in one, on which the publicized data are gathered in a column built in the provincial governmental service network, and hence there is no data set on the home page.

(2)   Dynamic data display on platforms

The home page of the data platform of Beijing shows the latest data sets and hottest download modules. The data platform of Shenzhen covers the latest data, hottest data, dynamic publicity, and latest application, as is shown in Fig. 4.4. In the five provinces and municipalities, Shenzhen has four dynamic display entries, ranking first. Guizhou, Shanghai, Beijing, and Zhejiang have three dynamic display entries. Through the dynamic display, the user can see the data sets that they are looking for and download them.

(3)   Visualization display on platforms

Data publicity platforms of Beijing, Shanghai, Zhejiang, Guizhou, and Shenzhen provide two visualization displays, namely, the display of data mapping of relevance

	Province/Municipality	Data publicity platform
Table 4.1 Websites of the five provincial or municipal data publicity platforms	Beijing	www.bjdata.gov.cn
	Shanghai	www.datash.gov.cn
	Zhejiang	data.zjzwfw.gov.cn
	Guizhou	www.gzdata.gov.cn
	Shenzhen	opendata.sz.gov.cn

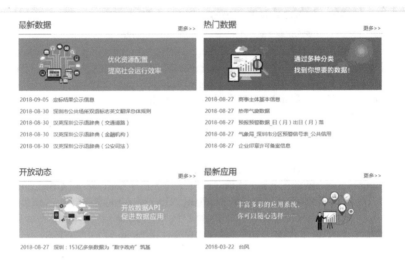

Fig. 4.4 Dynamic display on the data publicity platform of Shenzhen

between platform data and the display of platform data publicity and browsing, for instance, the departments and topics with the most publicized data. The platforms of Beijing and Zhejiang do not have visualization display modules on them, while those of Shanghai, Guizhou, and Shenzhen display data publicity and browsing. For instance, as is shown in Fig. 4.5, the platform of Shenzhen provides interface renewal, catalog renewal, and word cloud. The visualization display of platform data can help the user to know about the data on the platform more visually.

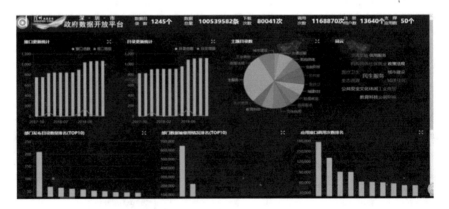

Fig. 4.5 Data publicity platform of Shenzhen

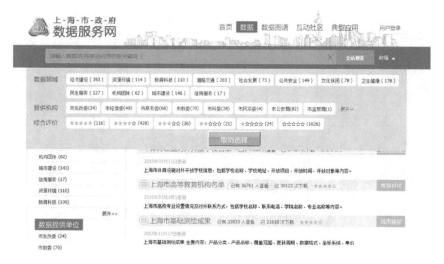

Fig. 4.6   Convenient searching on the data platform of Shanghai municipality

(4)   Convenient searching on platforms

Data publicity platforms of the five provinces and municipalities all provide convenient searching. Certain data sets can be found through key words. Those of Beijing and Shenzhen municipalities provide advanced searching. That is, more accurate searching can be done through the setting of more selection conditions. All local governmental data publicity platforms provide multiple data classification searching and navigation ways, for example, by topic and institution. In the searching and navigation system of Shanghai Municipality, comprehensive rating is adopted so that the users can search for the information in accordance with rating, as is shown in Fig. 4.6.

(5)   Interaction and communication on platforms

Data publicity platforms of five provinces and municipalities all provide interaction and communication functions concerning social involvement and public information publicity. On the data platforms of Shanghai Municipality and Guizhou Province, two ways of star rating and literal rating are adopted to show the users' scoring and comments on data sets and allowing users to evaluate data sets without registration. Figure 4.7 shows the evaluation function of the data platform of Shanghai Municipality. The interaction and communication function enhances the interaction between the platform and the user, and between users. The platform can respond in time to the users' demand for data and improve data quality.

(6)   Tool service function of platforms

Tool service function of platforms covers the basic tools provided by governmental data publicity platforms to help users to analyze and develop data sets. In the five

Fig. 4.7 Evaluation function of Shanghai municipal data platform

provinces and municipalities mentioned above, Guizhou, Shanghai, and Beijing provide some tools, including visualization analysis and development tools.

2. **Defined opened data scope**

Among the five provinces and municipalities, local governments have launched some development of information resources of different fields, including economic construction, resource environment, education technology, road transportation and communication, and people's livelihood. Beijing and Guizhou had the most data areas on their data platforms, about 20 of them (Table 4.2).

The number of data sets and publicity departments on data publicity platforms in the five provinces and municipalities by October 2018 is shown in Figs. 4.8 and 4.9. Shanghai Municipality has the most data resources for economic construction in data areas, with total 392 data sets; Shenzhen Municipality has the most data resources for livelihood service, with 638 data sets;Beijing Municipality also has many economic construction data resources, with 298 data sets.

3. **Improved quality of publicized data**

To improve the quality of publicized data, we have to improve the data's completeness, accuracy, validity, timeliness, and machine readability. In particular, the last one is vital to the improvement of quality. In the five provinces and municipalities, all data sets can be downloaded through machine reading, such as the CSV and XLS formats. The download format for the data of the five provinces and municipalities supports machine reading. The machine-readable data of Guizhou accounts for 96.75%, and that of Shenzhen, 96.27%.

Table 4.2 Opened data entries and resource quantity of the five provinces and municipalities

Provinces and municipalities	Data entries (quantity in brackets)
Beijing	Economic construction (298), sports and recreation (124), education and research (81), business service (75), social security (71), governmental institutions and social organizations (65), environment and resource protection (60), medical care and health (51), livelihood service (47), tourism and accommodation (40), transportation service (34), finance and taxation (27), agriculture and rural areas (25), job and employment (19), credit service (15), foods and drinks (8), living safety (8), consumption and shopping (7), religions and beliefs (7), and housing (4)
Shanghai	Economic construction (392), road transportation and communication (189), health (175), public security (147), urban construction (143), livelihood service (124), environment and resource protection (110), education and research (109), culture and recreation (77), social development (67), institutions and organizations (62), and credit service (17)
Provinces and Municipalities	Data entries (with the number of data sets in brackets)
Shenzhen	Livelihood service (638), culture and recreation (163), institutions and organizations (161), education and technology (130), medical care and health (126), public security (123), urban construction and housing (118), ecological resources (111), industry and trade (73), transportation and communication (59), credit service (48), finance and taxation (33), policies and laws and regulations (7), and urban construction (1)
Guizhou	Finance and taxation (172), transportation and communication (38), medical care and health (28), resource and energy (56), technical innovation (40), livelihood service (368), institutions and organizations (517), industry and agriculture (77), social security and employment (6), ecology and environment (13), business and trade (10), legal service (85), secure production (30), weather service (5), education and culture (83), geography and space (17), public security (33), credit service (7), urban construction and housing (22), and market supervision (13)
Zhejiang	Economic construction (134), resource and environment (30), urban construction (25), road transportation (16), education and technology (12), culture and recreation (30), livelihood service (21), and institutions and organizations (32)

For the convenience of the public in obtaining and developing data, all the provinces and municipalities mentioned above provide data interfaces and applications. The proportion of data sets that can be downloaded through API is shown in Table 4.3, all being no less than 30% as required by *Work Plan for the Pilot Program of Opening of Public Information Resources*.

The data publicity platforms of the five provinces and municipalities provide evaluation and error correction functions to meet the users' demand for and suggestions on data, enhance the publicized data examination and renewal and hence guarantee the data's completeness and timeliness. The data publicity platforms of the five

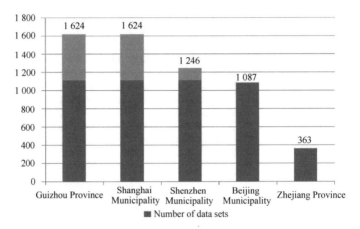

Fig. 4.8 Number of data sets on the data publicity platforms of the five provinces and municipalities

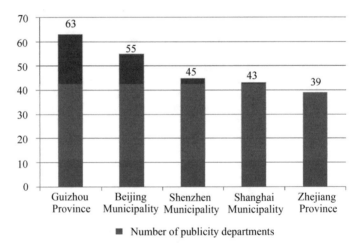

Fig. 4.9 Number of publicity departments on the data publicity platforms of the five provinces and municipalities

Table 4.3 Proportion of data sets that can be downloaded through API of the five provinces and municipalities

Provinces and municipalities	Proportion (%)
Beijing	100
Shanghai	39.58
Guizhou	55.23
Zhejiang	62.80
Shenzhen	79.54

provinces and municipalities show that the latest data of four of them has been dated to September 2018.

4. **Deepened use of publicized data**

The applications of publicized governmental data can be categorized into three kinds, namely, APPs, transmission products, and research results. The five provinces and municipalities mainly adopt APPs, and only the column named Endless Data opened by Guizhou Province is a transmission product. There are more and more areas and topics of the opened data APPs on the governmental platforms of Shanghai and Guizhou, covering livelihood service, economic construction, social development, weather and environmental protection, transportation and communication, health, and public security.

5. **Improved institutions and regulations**

The five provinces and municipalities have issued or launched policies to promote public information resource publicity, as is seen in Table 4.4. In particular, the provinces and municipalities listed as pilot regions in 2018 in that respect have

Table 4.4 Some of public information resource publicity policies of the five provinces and municipalities

Provinces and municipalities	Policies	Issuing time
Beijing	Measures for administration of Beijing governmental information resources (Trial)	12/2017
	Development and action plan for Beijing's big data and cloud computing (2016–2020)	08/2016
Shanghai	Work plan of the year 2017 for Shanghai governmental data resource sharing and publicity	08/2017
	Work plan of the year 2016 for Shanghai governmental data resource sharing and publicity	05/2016
Zhejiang	Work plan for Zhejiang's deepening "Internet + Governmental Service"	05/2017
	Measures for Zhejiang public data and e-government	03/2017
Guizhou	Implementation measures of Guiyang governmental data sharing and publicity	03/2018
	Rules of Guiyang governmental data sharing and publicity	04/2017
Shenzhen	Development and action plan for Shenzhen's promotion of big data (2016–2018)	10/2016
	Key points of work plan for Shenzhen's reform of promoting Internet + Governmental Service in 2017	03/2017

Table 4.5 Policies adopted by the five provinces and municipalities to guarantee the security of publicized information

Provinces and municipalities	Policies	Issuing time
Beijing	Beijing municipal contingency plan on cyber and information security	06/2013
Shanghai	Shanghai municipal action plan on industrial control system information security (2018–2020)	09/2018
	Measures for Shanghai classified information security protection evaluation organization management	10/2013
Zhejiang	Rules of Zhejiang classified information security protection records	01/2015
Guizhou	Guizhou water resources department's contingency plan on cyber and information security	11/2016
Shenzhen	Shenzhen civil affairs administration's contingency plan on cyber and information security	09/2016

made more detailed policies. For instance, Guiyang of Guizhou Province has issued *Implementation Measures for Guiyang Municipal Governmental Data Sharing and Publicity.*

6. **Reinforced publicized data security**

Local governments have reinforced the publicized data security, established relevant management institutions and confidentiality censorship, and taken all measures to ensure the safety of the data. They have set up and improved the emergency-responding mechanisms and contingency plans, encouraging all regions to take feasible security measures and innovate assessment means to guarantee coordinated development of public information opening, information security, and public interests. The five provinces and municipalities have all launched policies to enhance the security of publicized information, as is shown in Table 4.5.

4.5.3 Initial Effect of Public Information Publicity

1. **Improved governmental capability and level**

Through the development, exploration, and utilization of open public information, the government can efficiently improve their governance capability. For instance, Guizhou has launched the Iron Cage of Data and Intelligent Fire Prevention projects. The former is a practice of the Guiyang Government of Guizhou in applying publicized data to modern governance. Through departmental data collection, integration, analysis, and application, the project contributes to the construction of an iron wall for risk control and construction of a clean and honest government. A totally new mechanism has been established, with data playing its key role in decision-making, management and innovation. Data during the governmental administration is integrated and analyzed so that potential hazards can be found and controlled, and hence the power can be confined to the institutional cage and the governance capability can be improved. Intelligent Fire Prevention was launched in March 2016 as a pilot project. Data from 119 fire alarm hotline, government, and industrial sectors is integrated to provide support for accurate scheduling, quick police officer sending, and scientific handling, so that fire prevention can be commanded in real time and affairs can be handled by fire prevention armies.

2. **Facilitated vitality of traditional industries and development of emerging industries**

Public information publicity is beneficial to releasing the value of data, providing impetus for the development of digital economy, and facilitating the vitality of relevant industries. First, it boosts the integrated development of digital and real economy and promotes traditional industries to develop toward digitalization, networking, and intelligence. For instance, Shanghai Municipal Government has proposed the construction of digital water resource administration, having launched the country's first mobile phone inquiry software for water supply, namely, "My Tap Water", which enables the public to be clear about water quality. Its "scheduled water failure" and "scheduled meter change" functions help the public to arrange their daily life in advance to reduce the effect of water cutoff and water meter change and improve the service of the traditional industry. Secondly, public information publicity facilitates the combination between big data and cloud computing, IoT and AI, and brings about new industries and new modes of the digital economy. It has brought about a series of new industries with a focus on information integration and analysis, such as Qichacha, Tianyancha, and Qixin, which provide multiple information services for businesses and individuals by integrating and exploring public information resources.

3. **Improved public service**

Data publicity in the areas of education, medical care, employment, social security, housing, and communications can facilitate the development of "Internet + education" and "Internet + medical care", and hence improve the equality, universalization, and convenience of public service and meet the demand of the people for a better life. For instance, the platform named Shi'ance covers the introduction, pictures, features, ingredients, certifications, self-examination reports, inspection reports, sampling reports and nutrients of 22 food categories, 38 food subcategories, and 56 types of food, all stipulated by China Food and Drug Administration, so that consumers can learn about the food security to guarantee their diet health.

Chapter 5
Cleaner Cyberspace

5.1 Overview

In 2018, China continues to strengthen cyber content building, and improve the comprehensive cyber administration. Network media are thriving and the cyberspace is becoming cleaner and cleaner.

1. **Continuously enhanced positive publicity on the Internet**. Innovations have been made in the publicity and report of important meetings, events and decisions. News products are made and distributed, news resources are integrated and transmission is facilitated through the combination of AI, AR, biosensing and data mining. The List of China's News Website Communication Capacities shows that people.cn, xinhuanet.com and cctv.com take the lead in news websites. Websites like eastday.com, cqnews.net and rednet.cn are on the top of provincial news websites.
2. **Continued network communication base construction and team management**. There are altogether 343 Internet news information service organizations in China, 19, 562 governmental operation websites and 134, 827 weibo accounts of governmental organizations approved by sina.com. Key theory websites represented by qstheory.cn and gmw.cn have seen progress in their construction.
3. **Increasingly thriving network culture and successful "going overseas" of high-quality culture**. Nearly 200 network literary works have been authorized to be downloaded by overseas Internet users. There are over 100 overseas websites translating and spreading Chinese network literary works to other languages and countries. In 2017, online games independently developed by China saw a sales revenue of 8.28 billion US dollars, with a year-on-year increase of 14.5%.
4. **Continuously improved comprehensive governance of the network**. Laws and regulations have been launched concerning content management, technical security, e-commerce, and personal information protection. Concentrated control has been exercised over online live streaming, network reproduction, short video,

online games and online cartoons. China's Internet United Rumor Refuting Platform has been launched, and China's Network Social Organization Association has been founded, with 300 members at the beginning.

5. **Steadily expanded and deepened media integration**. Mainstream media are enhancing their new media construction. The new pattern of network communication with the active participation of commercial websites and professional vertical websites has been basically formed, and the construction of county-level convergence media is being accelerated.

6. **Thriving new forms of industry**. Network news APP has become the second largest basic APP, coming after instant messaging. Online live streaming is developing steadily, and voice live stream is a highlight while short video enjoys vitality. New media investment and fund-raising are active, and paid content industry is thriving.

5.2   Continuously Enhanced Network Content Construction

The report at the 19th CPC National Congress made it explicit that "cyber content building should be enhanced". Online publicity and reporting should be based on technical reform, so that the communication capacity, leadership, influence and credibility of news media can be improved steadily. Major Websites will be built and a number of high-quality online cultural products will be cultivated. Content construction, special rectification and team management will be done simultaneously to make the cyberspace cleaner.

5.2.1   Steadily Increasing Influence of Positive Publicity

The influence of the publicity of online topics, important meetings and events is being improved steadily, and innovation in transmission content, media forms, production technology and distribution channels keeps being done.

The publicity and interpretation of the following is prevailing the cyberspace, which sees positive energy and mainstream voice on it. They are Xi Jinping Thought on Socialism with Chinese Characteristics for a New Era, and the 19th CPC National Congress, the National People's Congress and the Chinese People's Political Consultative Conference ("Two Conferences" for short hereinafter), the 200th anniversary of Marx's birthday, the fortieth anniversary of China's reform and opening-up, the 2018 annual conference of Boao Forum for Asia, Shanghai Cooperation Qingdao Summit, and Beijing Summit of Forum for China–Africa Cooperation. Network news media has launched large-scale exhibitions and thematic reports on great events of China in various forms, including micro-video, online survey, H5 products, short video and so on. xinhuanet.com has launched the Theory Academy, which covers columns like "Theories in 100s", "40 Years' Stories in Pictures' and "Experts' Comments", for all of which theoretical convergence media have been adopted to achieve the best results. On people.cn, Two Conferences in Rap Cartoons throughout the country had 20 million visits, and over 42,000 users forwarded, commented and gave thumbs up to the column.[1] *Reading the Report beyond Paper*, a creative video, showed China's achievements in the past five years as written in the report at the 19th CPC National Congress through the combination 3D pictures and Origami animation, harvesting 105 million visits.[2] In cooperation with *People's Daily* and some other media, peo ple.cn offered a live show On-going Two Conferences at the PC terminal, mobile terminal and its own APPs. For instance, Henan Time of On-going Two Conferences was broadcast live on March 15, 2018 simultaneously at the APP of people.cn, all CPC public information platforms, Tencent News, hnr.cn and The Moment.

The coverage of news has been expanded thanks to the integration of media resources and their cooperation with commercial platforms. By September 2018, *People's Daily* had had 20,000 e-reading columns, which have formed the largest interaction platform with the largest number of media.[3] CCTV broadcast the Two Conferences in cooperation with all APPs and over 50 WeChat and Weibo platforms. Only cctv.com itself issued 156 million reports visited for 800 million times, and its direct video broadcast saw 2.05 billion visits. First Report and Witnessing the 19th CPC National Congress, two series of short videos with elaborately edited CCTV content, publicized ten sessions with great results. The column "Bai Yansong, observer of the 19th CPC National Congress, Cheers for the New Era" was broadcast for five million times.[4]

As for technical innovation, AI, AR, biosensing and data mining are integrated into media. people.cn launched the online map depot system, with which pictures can be uploaded when they are taken. That improved the timeliness of Two Conferences reporting. Besides, an AI robot named Wangzai was introduced into the interview programs about Two Conferences. AR was adopted for the first time in reporting Two Conferences, for instance, Two Conferences with AR: people's livelihood and welfare. xinhuanet.com used Star biosensor to produce the country's first SGC, illustrating the emotional curve of the audience and pioneering in deep data exploration and emotional analysis, which could not be done by traditional reporting.

---

[1] http://www.cac.gov.cn & people.cn, *Simply Profound Innovation*, http://www.cac.gov.cn/2018-07/30/c_1123195166.htm.

[2] xinhuanet.com, Original Innovative Short Video Named *Reading the Report beyond Paper*, http://www.xinhuanet.com/company/2018-03/12/c_129827923.htm.

[3] XU Tao, *Three Directions of the Expasion of Major Business*, http://media.people.com.cn/n1/2018/0910/c14677-30284407.html.

[4] *Safe, Accurate and Colourful Broadcasting of the Two Conferences by CCTV in 2018*, http://www.cctv.com/2018/03/26/ARTIOVRgZNSBi02eTgUCauIJ180326.shtml.

5.2.2   Capacity Building in Communication, Leading Power, Influence and Credibility Being the Key Goal of News Media

Internet news information service organizations cooperate with each other in traffic and content to improve their communication capacity, leading power, influence and credibility. Central news websites have adopted orientation guidance, channel expansion, process reproduction, organizational optimization and system and mechanism reform to enhance content building and deepen integration of media, playing the role of the wind vane. Local news websites have been rooted in the grassroots, serving the masses of people. The potential of local cultural transmission has been tapped, showing regional and public service capabilities. Commercial websites continue to be the gateway of the transfer between new and old media, adopting innovation for more transmission space.

In September 2018, the journal *New Media* launched the List of Transmission Capacity of China's News Websites, made with communication capacity, leading power, influence and credibility proposed by President Xi Jinping as core indicators, to guide all news websites in their news and public opinion promotion.

In the first half of 2018, people.cn, xinhuanet.com and cctv.com were the top three in comprehensive communication capability, as the authoritative sources of news information. They play a leading and cohesive role in that respect. cctv.com keeps increasing in the ranking in the streaming media era (Table 5.1).

In the first half of 2018, eastday.com (Shanghai), cqnews.com (Chongqing) and rednet.cn (Hunan) were the top three on the list of comprehensive communication capacity of major provincial news websites. Those top 30 scored over 85 (see Table 5.2), which reflects provincial news websites' innovation and achievements in promoting media integration, in covering Weibo and WeChat and APPs and in providing localized convenient service.

5.2.3   Continuous Construction of Media Platforms

There are more and more news platforms including mainstream websites, APPs, forums, blogs, microblogs, public accounts, instant messaging and online live streaming. By August 31, 2018, there were 343 Internet news service organizations, covering 351 websites, 264 APPs, 85 forums, 22 blogs, 3 microblogs, 1026 public accounts, instant messaging tools, 7 live streams and 8 other forms, providing 1767 services.[5]

Governmental website and new governmental media construction is going on. By October 2018, there were 19,562 governmental websites. New governmental media

---

[5]http://www.cac.gov.cn/2018-09/10/c_1122842142.htm.

Table 5.1 Comprehensive communication capacity list of major central news websites in the first half of 2018

Ranking	Websites	Comprehensive communication capacity index
1	people.cn	92.00
2	xinhuanet.com	84.19
3	cctv.com	83.67
4	chinanews.com	82.96
5	cyol.com	80.79
6	people.cn	80.48
7	china.com.cn	77.53
8	youth.cn	76.40
9	81.cn	76.17
10	ce.cn	76.07
11	gmw.cn	75.86
12	chinadaily.cm.cn	74.49
13	cri.cn	71.76
14	taiwan.cn	71.39
15	tibet.cn	70.00

WeChat Account of New Media, *Comprehensive Communication Capacity List of Major Central News Websites in the First Half of 2018*, https://mp.weixin.qq.com/s?__biz=MzA5OTA2O TU5Mg==&mid=2651479536&idx=1&sn=f43b3cd3fb6842bc9c3deae88250257e&chksm=8b7 90607bc0e8f116db36384d9d4bca850eea1ff8126736c621d0e90f187748b9b0b644f39dc&scene= 21#wechat_redirect

are the new channels for communicating society and the public.[6] Take the legal system as an example. By the end of 2017, 28,000 legal weibo accounts had been certified, together with 5600 public WeChat accounts and about 18,000 toutiao accounts.[7] The new governmental media matrix has been formed, composed of central, local and departmental sectors. Jiangning Public Security, For the Security of Beijing, Henan Procuratorate and Shandong Supreme Court are popular accounts and typical brands in terms of governmental information opening-up and governmental affairs publicity.

The China Media Group was founded to promote the integrated development of radio and TV media and emerging media. According to *the Decision of the CPC Central Committee on Deepening the Reform of Party and Government Institutions*, the former CCTV (CGTN), former China National Radio and former China Radio International were merged into the China Media Group, which plays an important role in enhancing the CPC's construction and administration of public opinion platforms and the overall strength and competitiveness of radio and TV media

---

[6]http://114.55.181.28/databaseInfo/index.

[7]Public Opinion Center of gmw.cn, *Blue Paper of New Media APPs of the Legal System* 92017) [EB/OL], http://legal.gmw.cn/2018-06/15/content_29296054.htm, June 15, 2018.

Table 5.2 Comprehensive communication capacity list of major provincial news websites in the first half of 2018

Ranking	Websites	Comprehensive communication capacity index
1	eastday.com	92.76
2	cqnews.com	92.46
3	rednet.cn	92.41
4	dawww.com	92.20
5	thepaper.cn	91.76
6	zjol.com.cn	91.69
7	jxnews.com.cn	90.89
8	southcn.com	90.83
9	iqilu.com	90.49
10	newssc.org	90.29
11	jiemian.com	90.17
12	kankanews.com	89.81
13	ycwb.com	89.69
14	qianlong.com	89.42
15	gog.cn	89.38
16	shobserver.com	89.12
17	enorth.com.cn	89.06
18	dahe.cn	88.58
19	gdtv.cn	88.53
20	cnwest.com	88.17
21	hinews.cn	87.99
22	hebei.com.cn	87.88
23	voc.com.cn	87.55
24	thecover.cn	87.07
25	jschina.com.cn	86.88
26	gansudaily.com.cn	86.87
27	sxgov.cn	86.84
28	sdnews.com.cn	86.45
29	anhuinews.com	86.11
30	mgtv.com	85.84

WeChat Account of New Media, Comprehensive Communication Capacity list of Major Provincial News Websites in the First Half of 2018, https://mp.weixin.qq.com/s?__biz=MzA5OTA2O TU5Mg==&mid=2651479536&idx=2&sn=d74ad56d7620dfef8ce2f3e72e47f155&chksm=8b7 90607bc0e8f11e946ce750fab3aad12c13395fc904e80814225017ee3fcb896eed78c8b8f&scene= 21#wechat_redirect and emerging media, accelerating international transmission capacity-building, and communicating China's voice to the world.

China's communication capacity is being reinforced. *People's Daily* and Xinhua News Agency see the number of their fans on Facebook and Twitter increasing. The number of the fans of CCTV news on Facebook has reached more than 30 million. By the beginning of 2017, CCTV.com had opened on overseas mainstream platforms 31 accounts, including the CCTV series and Panda Channel series in Chinese, English, Spanish, French, Arabic and Russian. In October 2017, the English APP of *People's Daily* was launched, providing broadcasting to overseas users in English in the form of mobile interconnected platforms. In November 2017, new media of the former CGTN enjoyed over 300 million clicks on YouTube, the world's largest video website, and the total watching time length reached 370 million minutes.[8] In January 2018, the English APP of Xinhua News Agency was launched as a new channel of international communication. The English APP of *China Daily* is an English news APP attracting 10 million clicks, with its users from over 140 countries and regions. *China Daily* has 53 million users on Facebook.[9]

5.2.4 Increasingly Thriving Internet Culture

There are flourishing positive-energy online products. In July 2018, the positive-energy online products selection was held, with the theme "positive energy on the Internet and a new era." This event highlighted excellent figures and products concerning great policies, topics, events, activities and hot issues and emergencies as well as the achievements and progresses in positive energy and new media development in the past year.[10]

Traditional excellent culture of China is being rejuvenated in the era of the Internet. In 2018, the country carried out the project of the Spring Festival's going out. For that, a series brand cultural events were launched, including "Merry Spring Festival", "Spring Festival Celebration from All Directions" and "Global Lantern Festival", and a number of film and TV programs, books, cartoons and cultural creation products were promoted through network platforms with a number of fans. In September 2017, large-scale network communication activities were started about non-tangible cultural heritage named "Welcoming the 19th National Congress of CPC: Eulogy to China" and "Non-tangible Cultural Heritage". Nearly 100 central and local news websites and commercial websites and 200 journalists from new cultural media and

---

[8] Over 300 million clicks of CGTN on YouTube, http://www.cctv.com/2017/11/09/ARTIgBXem qqcyU2b80mBWLwu171109.shtml.

[9] Tang Xujun, Huang Chuxin and Wang Dan, Smart Interconnection and Digital China: a new era of China's new media development, *China's New Media Development Report* (2018), Beijing: Social Sciences Academic Press (China).

[10] People.cn, Striving in the New Era: Five "One hundred" Allows the Smoothness of Positive Energy, http://politics.people.com.cn/n1/2018/0717/c1001-30153432.html.

traditional media reported the inheritance and development of non-tangible cultural heritage, demonstrating the unique charm of excellent traditional culture.

Cultural products on the Internet are gaining an increasing popularity in the world. In the past year, nearly 200 Chinese Internet literary works have been authorized to be downloaded by overseas users. Some Internet movie and TV products adapted from those works are popular overseas. By March 2018, there were about 100 overseas websites providing translations in different languages and spread of Internet literary works of China. Quite a few Internet works have gained overseas copyright and have been translated into English, Japanese, Korean and Vietnamese. More and more Chinese Internet literary works have been introduced abroad through overseas merging, joint operation and establishment of branches. Chinese games are producing increasing influence and gaining increasing position in the overseas market. *China's Game Industry Report 2017* shows that, in the year 2017, online games independently developed by China saw a sales revenue of 8.28 billion US dollars, with a year-on-year growth of 14.5%.

5.3   Continuously Improved Comprehensive Governance of the Network

At the National Cybersecurity and Informationization Working Conference, General Secretary Xi Jinping stressed that the comprehensive cyberspace governance should be improved, with CPC committees as the leader, governments as the manager, businesses as the performer, society as the supervisor, and Internet users as self-discipliner, so that a comprehensive cyberspace governance pattern can be formed by the combination of economic, legal and technical means.

5.3.1   Accelerating Legislation on Cyber Content

China attaches importance to legislation on the Internet, having issued a number of laws and documents concerning cyber content. For instance, in August 2017, Cyber Administration of China issued *Provisions on the Administration of Internet Comments Posting Services*; in September of the same year, it issued *Rules of Regulating Internet Group Information Services*; in October, it issued *Measures for the Administration of Content Management Practitioners in Internet News Information Service Providers* to strengthen administration of content management practitioners in Internet news information service, safeguard legal rights and interests of the practitioners and the public, and promote healthy development of Internet news information service. On February 2, 2018, Cyber Administration of China launched *Provisions on the Administration of Micro-blog Information Service*, which clarifies the accountability and identity certification of micro-bloggers, rumor-refuting mechanism, self-discipline in that industry, and social supervision and administration over micro-blog. In March 2018, the State Administration of Press, Publication, Radio, Film and Television issued the *Notice of Further Regulating the Communication Order of Online Video and Audio Programs*, in which special provisions are made for online movie clips, previews and video and audio program adaptations.

5.3.2   Special Actions Launched

Competent authorities cooperate with each other in launching special actions against the illegal actions online harmful to the people's life and social harmony, investigating into all cases and rectifying the order in cyberspace. In the first half of 2018, Ministry of Culture and Tourism inspected online performances and game markets. Law enforcement examination has been done concerning online show and game markets as well as show APP communication channels, and online show and online games with forbidden content. In August 2018, National Office against Pornography and Illegal Publications, in cooperation with Ministry of Industry and Information Technology, Ministry of Public Security, Ministry of Culture and Tourism, National Radio and Television Administration and Cyber Administration of China, issued *Provisions on the Administration of Internet Live-Streaming Services*, according to which the licensing and archive recording of Internet live-streaming service is reinforced, together with the basic administration of that respect, a long-term supervision mechanism on it is enhanced and illegal Internet living streaming services are wiped out. Sword Net 2018 was launched to rectify the copyright administration of network reproduction, short video and cartoons and to safeguard the order of copyright communication in terms of Internet live streaming, knowledge sharing and audio books.

5.3.3   Defining Businesses' Accountability

Businesses' accountability keeps being enhanced. Through platforms, illegal content is forbidden and users' actions are disciplined to keep the cyberspace clean. In April 2018, Cyber Administration of China and National Radio and Television Administration arranged talks with the heads of Kuaishou and Toutiao about the Under-age Mums, ordering them to rectify their short video platforms, including the forbidding of new audio and video program account uploading, examination of all existing accounts and strengthening of accountability of website verifiers and chief managers for law breaching and harmful programs. The two platforms have strengthened their self-discipline and rectified themselves by apologizing to the public, closed down some accounts and reinforced the examination.

5.3.4 Network Social Organizations and Industrial Self-discipline Organizations Playing Their Role

Network social organizations and industrial self-discipline organizations play their role in cyberspace ecology building and comprehensive cyber governance capacity improvement. Multiple participants interact and coordinate with each other for building clean cyberspace.

China Federation of Internet Societies was founded on May 9, 2018, with 300 initial members. It is a national united hub-like social organization, playing an active role in mobilizing Internet businesses to get involved in comprehensive cyber governance, safeguarding the Internet industrial order, developing public charities on the Internet, participating in the protection of cybersecurity, promoting informatization and digital economy and taking part in foreign communication and cooperation in terms of the Internet.

To control rumors on the Internet, the Law Breaching and Malicious Information Center of Cyber Administration of China, together with 27 organizations of Party School of the General Committee of CPC and National Development and Reform Commission, established China's Internet United Rumor Refuting Platform, which is hosted by xinhuanet.com. Key central websites and local refuting platforms, portal websites and think tanks all participate in it, providing authoritative suggestions on the identification and tipping-off of rumors.

Column 4: China's Internet United Rumor Refuting Platform

On August 29, 2018, under the auspice of the Law Breaching and Malicious Information Center of Cyber Administration of China, the country's Internet United Rumor Refuting Platform was launched on the Internet, hosted by xin huanet.com. It is a great action for controlling rumors on the Internet and fostering the cleanness of cyberspace, aimed to help the public to identify and tip off rumors.

Columns like "Announcement by Ministries", "Local Response", "Media Confirmation", "Expert's Views" and "Rumor Refuting Corner" have been set up on the platform, so that the public can tip off and confirm rumors, and obtain rumor refuting information provided by competent authorities and experts. Rumors can be identified and refuted in united authority, and information can be spread at multiple terminals. Users can click for confirmation, and prevention of rumors can be done in time.

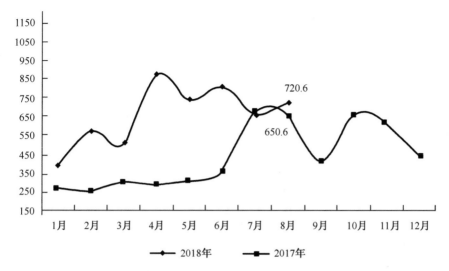

Fig. 5.1 Valid reporting and handling of bad and harmful information throughout China (Unit: 10 thousand)

5.3.5 Internet Users Being a Major Force for Cyberspace Governance

Internet users can be taken as a basic force for cyberspace governance. It is necessary to improve their quality and guide them in participating in social governance of the Internet, in order to safeguard cyber security. During the National Cybersecurity Publicity Week of 2018, the forum named "Co-building, Co-governance and Sharing: Internet users 'participation in safeguarding clean cyberspace" was held, aimed to promote the quality education of Internet users concerning cyberspace. In August 2010, 7.206 million cases concerning cyber security were reported (see Fig. 5.1), with the month-on-month and year-on-year growth rate respectively at 7.1 and 10.8%. The Law Breaching and Malicious Information Center of Cyber Administration of China handled 102,000 cases, with the month-on-month and year-on-year growth rate respectively at 29.8 and 22.1%; local cyber administration authorities handled 1.394 million cases, with the month-on-month growth rate at 25.6%, and the year-on-year decrease rate at 65.9%. Major websites handled 5.71 million cases, with the month-on-month and year-on-year growth rate at 1.4 times 10.8% respectively.[11]

---

[11]Law Breaching and Malicious Information Center of Cyber Administration of China, *Reporting and Handling of Bad and Harmful Information by August 2018*, http://www.12377.cn/txt/2018-09/12/content_40500194.htm.

5.4 Thriving Network Media

5.4.1 Steady Media Integration

Traditional media are being integrated and innovated. *People's Daily* is dedicated to promoting media integration, having succeeded in full-media platform integration of news production and planning, collection, editing and publicizing and set up the full-media "central kitchen" supporting quality content. The APP of *People's Daily* has been downloaded for 248 million times, its legal person's Weibo has 110 million fans, and its WeChat account attracts over 20 million, ranking among the top in mainstream media. people.cn has 258 million users.[12]

Convergence media centers at the country-level have emerged. From September 20 to 21, 2018, the Central Public Relations Department held a meeting on the construction of county-level convergence media centers. 600 such centers will be built in the coming year, and by 2020, they will be available throughout the country. This work will help to transfer media integration from the central and provincial-level to grassroots-level, from major ones to branch ones and the reform of the latter will facilitate the activity of the entire media system of the country. In July 2018, Liuyang Convergence Media Center was founded, and the first county-level CPC and governmental public relations platform of Shaanxi was launched.

Column 5: Guixi Convergence Media Center of Jiangxi Province

Guixi is a county-level city under the jurisdiction of Yingtan of Jiangxi. The Convergence Media Center there, which was established in October 2016, was ranked first in 100 counties (and municipalities and districts of the same level) of Jiangxi in terms of communication index of convergence media of Jiangxi in 2017. The "9 + 2" convergence media matric of Guixi is made up of the following new media and traditional ones: Guixi Announcement WeChat Platform, Guixi Daily WeChat Platform, *Jiangxi Mobile Phone Daily* APP, Guixi Hand APP, Guixi News Web, *Guixi Mobile Phone Daily* in Colors, Guixi Announcement Weibo, Guixi Announcement Toutiao Account, Guixi Douyin Account, and Guixi Daily and Guixi Radio and TV platforms.

The mode of "one collection, diversified production and multiple publi-cation" has been adopted in Guixi to integrate all media resources through new technologies. Integrated software has been developed, and the Internet Culture Society has been founded to promote the convergence of platforms. News supply channels are made smooth and platforms can share news materials and integrate the content. News communication is combined with culture and tourism, image demonstration, economic development and service for people's livelihood.

---

[12]Speech by Tuo Zhen, Editor-in-Chief of *People's Daily* at the Media Convergence Forum 2018, http://media.people.com.cn/n1/2018/0910/c14677-30282249.html.

Besides, integrated administration is exercised in Guixi, with the Municipal CPC Committee as the leader, all departments as cooperators and media as the coordinator. The Media Diagnosis Room has been set up to supervise the handling of the public opinions. The media team is made up of backbone forces and new media talents, with meetings held and correspondents trained regularly.

5.4.2 Steady Development of New Applications and New Industrial Forms

Applets enjoy fast development, having become the new hotspot of leading Internet businesses. WeChat and Alipay applets have been launched. Baidu smart applets and Toutiao applets are being tested. By the end of 2018, 1238 million WeChat applets had been launched, with 200 million active users, and over 1.5 million developers. Among those applets, there are 516 media ones, 1278 socializing ones, 297,000 store ones, 938,000 tool ones and 1278 game ones. Ant Financial Services Group has defined applet development as one of the strategies in the coming three years. On September 12, 2018, it launched Alipay applets, which totaled 20,900 in five days.

Online live streaming remains steady in development. In comparison with that in 2016, the number of online live streaming users increased slowly in 2017. By June 2018, China had 425 million online live streaming users, 2.94 million more in comparison with the number at the end of 2017, with the use rate of 53, 1.7% lower than that at the end of 2017.[13] The use rate of live sport shows increased, promoted by the World Cup in Russia in the half of 2018, in comparison with that at the end of 2017, but that of games, reality shows and vocal concert shows decreased a little (see Fig. 5.2). For the fast-developing online live streaming, competent authorities have strengthened administration over the live streaming content and platform quality, so that the industry can develop healthily.

Short video is developing rapidly and the number of users keeps increasing. From June 2017 to June 2018, the use time length of short video witnessed a year-on-year growth rate of three times. In particular, during the Spring Festival of 2018, short video APPs were introduced to third and fourth-tier cities, and the number of the users kept increasing. By June 2018, the total number of all hot short video users reached 594 million, accounting for 74.1% of the number of Internet users of China. The supervision system is being improved, guiding the key work towards ecological building, so short video industry is developing in a healthy, orderly and steady way.

---

[13] *Source* The 42nd China Statistical Report on Internet Development.

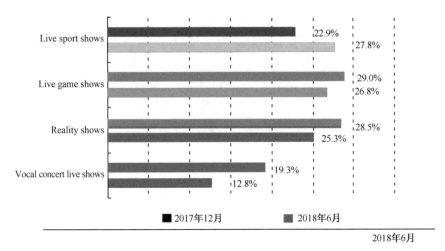

Fig. 5.2 Online live streaming use rate from December 2017 to June 2018 (*Source* CNNIC survey and statistics of the Internet development of China)

Quality content promotes the increase of the use rate of online audio (listening media). In the past year, audio APPs represented by Ximalaya FM, Qingting FM and Lizhi FM has seen their use rate increasing dramatically. With the thriving of their paid content, these platforms will see a new round of development.

Chapter 6
Steady Improvement of Cybersecurity Safeguarding Capacity

6.1 Overview

Since the 18th National Congress of CPC, the Central Leadership with Xi Jinping as the core has been attaching importance to cybersecurity, and has launched a series of measures and laws in terms of the construction of cybersecurity mechanisms and systems, industrial and technical development, cyber protection capacity improvement and talent production, and security awareness education. China's cybersecurity system construction is witnessing rapid development, its cybersecurity guarantee capacity building is being reinforced, and the national security shield is being consolidated.

(1) There are increasingly complicated problems in terms of cybersecurity, with more Advance Persistent Threats (APTs)and increasing Distributed Denial of Service (DDoS) peak value, and frequent online frauds, phishing emails, mining Trojans, and private information disclosure and export.

(2) Since the adoption of *Cybersecurity Law of the People's Republic of China*, the country has been promoting the top design of cybersecurity administration while accelerating facilitating legislation, carrying out cybersecurity guarantee, carrying out critical information infrastructure security examination, formulating cybersecurity standards, and launching industrial Internet security action plans.

(3) Cybersecurity industry is thriving. In 2017, the volume of cybersecurity industry reached 43.92 billion *yuan*, with 2,681 businesses involved, whose number has increased dramatically. The cybersecurity market is subdivided, and businesses' innovation awareness keeps increasing. Cybersecurity technology tends to be increasingly automatic and intelligent, and cloud and data security are seeing dramatic development.

(4) Cybersecurity serves and relies on the people. While accelerating the cultivation of cybersecurity talents, the country hosts the National Cybersecurity Publicity Week, a variety of cybersecurity meetings, and cybersecurity technology competitions to improve the people's cybersecurity awareness.

© Springer Nature Singapore Pte Ltd. 2020
Chinese Academy of Cyberspace Studies, *China Internet Development Report 2018*,
https://doi.org/10.1007/978-981-15-4043-1_6

6.2   Increasingly Complicated Cybersecurity

China is faced with increasingly complicated problems in terms of cybersecurity, with frequent ransom attacks, increasingly difficult DDoS peak value, threats to intelligent networked equipment, hazards for industrial control systems, and active APTs.

6.2.1   Serious Situation for Cybersecurity

CNCERT statistics show that in the first half of 2018, there were 1,138 industrial Internet devices, 1,968 industrial Internet web monitoring administration systems and cloud platforms, and 520 financial websites with high-risk vulnerabilities. In the same period, there were 413 new security vulnerabilities in the telecommunication sub-vulnerability banks and 261 new security vulnerabilities in the industrial control system vulnerability banks, accounting for 5.3 and 3.3%, having witnessed an increase in comparison with that of 2017. The critical information infrastructure security vulnerabilities include Cisco Smart Install distant order execution vulnerabilities,[1] CPU Spectre and Meltdown vulnerabilities,[2] and WebLogic Server WLS nuclear component deserialization vulnerabilities.[3]

6.2.2   More and More Security Problems Concerning Mobile Internet and IoT

With the rapid development and popularization of mobile Internet technology, there is an increasing percentage of Internet attacks on mobile terminal devices and smart devices of IoT. According to CNCERT statistics, in the first half of 2018, the number of malware programs attacking mobile Internet increased by about 1.48 million, accounting for 58.5% of the total increase in 2017. In the same period, 1,130 security vulnerabilities were recorded on CNVD, with a year-on-year increase of 17.3%, and there were about 17,000 malware program-controlled server IP addresses aimed to attack IoT devices, up 11.3% from that in the latter half of 2017, with about 85.3% of the IP addresses located overseas. The number of malware programs from 2010 to 2017 aimed to attack the mobile Internet is shown in Fig. 6.1.

---

[1]CNVD-2018-06774,corresponding to CVE-2018-0171.

[2]CNVD-2018-00303 and CNVD-2018-00304,corresponding to CVE-2017-5754 and CVE-2017-5753 respectively.

[3]CNVD-2018-07811,corresponding to CVE-2018-2628.

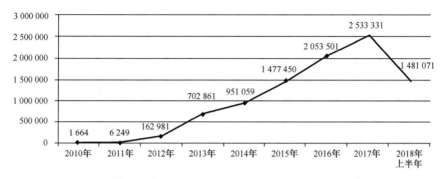

Fig. 6.1 Number of malware programs aimed to attack mobile Internet (2010–2017). *Source* CNCERT/CC

6.2.3 APT Aimed at Key Organizations of the Country

APT attacks pose a threat to national security, geographic politics, and economy. The areas subject to such threat are government, military forces, energy, aerospace, telecommunication, media, finance, and Internet, which are all key sectors. By the end of December 2017, 360 and Antiy detected 38 APT organizations aiming to attack China's Internet. The most active ones include Ocean Lotus and APT-C-09, and their frequent attacks include harpoon mail and water hole.[4] According to CNCERT statistics, in the first half of 2018, there were nearly 600 malicious codes and their extended documents spread through mails. Phishing mails can be categorized into those concerning orders, payment and notice, and their attachments include malicious vulnerability documents, executable files, and compression codes of malicious codes and malicious macro files.

6.2.4 Increasingly Difficult Defense Against DDoS

DDoS attacks are used in malicious competitions within an industry and in data thefts and other cybercrimes. Their number and peak value keep increasing year by year. Those above Tbps level have become the trend. Since February 21, 2018, Memcached reflection attacks have been active in China. At about 2 a.m., March 1, 2018, their peak value in the country reached as high as 1.94 Tbps.[5] More gangs are launching DDoS attacks, involving tool developers and other staff, with their labor divided into technology, sales, and attack.

---

[4]360 Threat Intelligence Center, *APT Research Report of China* (2017); Antiy, *Green Stain Action: research report on years of attacks*.

[5]CNCERT,*General Information of Memcached Server Reflecton Attacks.*

Column 6: DDoS Resources[6]

(1) Control terminal resources, referring to Trojan or botnet control terminals launching DDoS attacks by controlling Trojan or Zombie host nodes.

(2) Broiler resources, referring to the Zombie host nodes controlled by control terminals for DDoS attacks.

(3) Reflection server resources, referring to servers and hosts taken advantage of by hackers for reflection attacks. Some services not needing to be identified but having the effect of amplification and being easily deployed on the Internet (such as DNS servers and NTP servers) can easily become online resources for DDoS.

(4) Cross-domain fake traffic source routers, referring to routers that forward large amounts of fake IP attack traffic across domains. Since China requires that operators have their source address verified when accessing the Internet, the existence of cross-domain fake traffic indicates that there may be defects in the source address verification configuration of the router or the one below it, and that there may be devices launching DDoS attacks in the network of the router.

(5) Local fake traffic source routers, referring to the routers that forward large amounts of fake IP attack traffic within the domain. This indicates that there are devices launching DDoS attacks in the network of the routers.

6.2.5 Endless Emergence of Cyber Frauds

Traditional online frauds like malicious link sending and bank information phishing keep emerging. According to CNCERT statistics, in December 2017 there were 254 active phishing websites, and 3,388 IPs logging in such websites. 31.22% of IP users submitted their information when logging in, so they were the substantially affected people.[7] Meanwhile, new cyber frauds through socializing networks, APP software, and QR codes keep emerging. According to CNCERT statistics, in the first half of 2018, nearly 400,000 users were affected by such cyber frauds.

---

[6]CNCERT,*Analysis Report on China's DDoS Attack Resources in 2017.*

[7]CNCERT, *Early Warning and Analysis of Cyber Frauds on Phishing Websites.*

6.2.6 Frequent Emergence of Malicious Attacks like Ransom Attacks and Mining

In 2017, the price of digital currencies like Bitcoins and Ether coins saw a sharp increase, and attacks on digital currency transaction platforms frequently occurred, which gave rise to more ransom attacks on digital currencies and malicious programs of mining. Since 2018, there has been an increasing number of mining Trojans. There were 45 of them in the first half of 2018, more than the total of 2017.

6.3 Upgrading of Cybersecurity Mechanism

Without cybersecurity, there would be no national security. Prevention has to be done to guarantee safety. A proper outlook on cybersecurity should be formed to improve the cybersecurity protection capacity. Great progress has been made in top design, legislation, and service support in China, which has laid a solid foundation for cybersecurity administration.

6.3.1 Steady Progress in Top Design of Cybersecurity Administration

In March 2018, the Central Committee of CPC issued *Decision on Deepening Reform of Party and State Institutions*, and the Central Leading Group for Cyber Affairs was reorganized into the Central Committee for Cyber Affairs, who is in charge of top design, general layout, coordination, overall promotion, and implementation of important work. The National Computer Network and Information Security Administration Center which had been administered by Ministry of Industry and Information Technology became administered by Office of the Central Leading Group for Cyberspace Affairs. This has enhanced the coordination and administration capacity of cybersecurity.

6.3.2 Improved Legislation on Cybersecurity

Cybersecurity Law of the People's Republic of China is a fundamental law of the country in terms of cybersecurity. Since its adoption, some supporting laws and regulations have been promulgated or are being planned.

1.   **Importance is attached to personal information security and key data**

In April 2017, *Methods of Outbound Personal Information and Key Data Security Assessment (Exposure Draft)* was issued, followed by *Information Technology— Personal Information Security Specification* and relevant standards. Thanks to the documents, importance is attached to personal information security and data protection. The Action of Personal Information Protection Upgrading is implemented by Cyber Administration of China, Ministry of Industry and Information Technology, Ministry of Public Security and National Standardization Administration, and privacy provisions are put into effect in particular. Privacy provisions of ten online products and services including WeChat and Taobao have been reviewed for the first time. In December 2017, *Report of the Implementation of the Decision on Reinforcement of Online Information Protection* was reviewed at the 31st Conference of the Standing Committee of the National People's Congress. According to the report, China's personal information protection is faced with a serious situation, so it is required that the citizens' personal information protection system should be adopted in legal examinations and crime should be found out concerning infringement on personal information.

2.   **Standardized measures against cybersecurity threats**

National Emergency Response Plan for Internet Security Incidents defines and classifies cybersecurity incidents and contains provisions on early warning, emergency handling, investigation and assessment, and prevention guarantee of cybersecurity incidents. *Guide of Emergency Administration of Industrial Control System Information Security* proposes a series of requirements for industrial control security risk monitoring, information report and communication, and emergency handling and administration in sensitive periods, as well as for accountability, workflow, and guarantee measures. *Regulations on Cyber Security Threat Monitoring and Handling* issued in August 2017 defines cybersecurity threats and the accountability of competent authorities, professional institutions and businesses concerning the threat. *Contingency Plan for Internet Security Incidents* issued in November 2017 defines the organization system, incident classification, monitoring and early warning, and response concerning emergencies of Internet security incidents as well as prevention and emergency preparation and their corresponding guarantee measures.

3.   **More explicit requirements for online product and service security**

In June 2017, Cyberspace Administration of China, together with Ministry of Industry and Information Technology, Ministry of Public Security, and Certification and Accreditation Administration of the People's Republic of China, issued *Catalogue of Exclusive Products of Critical Internet Equipment and Cyber Security (First Batch)*, which defines the institutional qualifications of the certifier or monitor of exclusive products of critical Internet equipment and cybersecurity and requires that all monitoring results should be reported to competent authorities. National Information Security Standardization Technical Committee (NISSTC) has studied

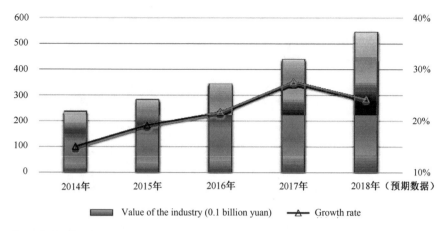

Fig. 6.2 Predicted value of China's cybersecurity industry in 2018. *Source* China Academy of Information and Communications Technology

and made general requirements for Internet product and service security and security detection, and hence improved relevant assessment systems.

6.4 Steady Development of Cybersecurity Industry and Technology

6.4.1 Development of Cybersecurity Industry

From 2017 to 2018, China's cybersecurity industry hit a new high in its scale. According to China Academy of Information and Communications Technology, the total value of that industry reached 43.92 billion *yuan* in 2017, with an increase of 27.6% in comparison with that in 2016, and it is expected to amount to 54.549 billion *yuan* in 2018 (see Fig. 6.2). According to statistics, there were 2,681 cyber-security businesses in China in 2017, most of them located in Beijing, Guangdong, and Shanghai, with the respective number of 957, 336, and 279 (see Fig. 6.1). In the same year, 189 new businesses emerged in that field (Table 6.1).

The year 2017 saw a stable performance of China's domestic listed security busi-nesses, whose revenue kept increasing for 3 years. Let's take the 10 listed ones as an example (see Fig. 6.3) to show their revenue and investment in R&D. Their average revenue in 2017 was 1.572 billion *yuan*, with a growth rate of 26.98%, above the average of their international counterparts, which is 21.99%. Their average net profit was 256 million *yuan*, which saw an increase of 22.68% in comparison with that in 2016. The total revenue of Venustech, Westone, Bluedon, Lanxum, and Sangfor was over two billion *yuan*. With incentive policies launched and demand for cyber-security increased, the growth rate of those businesses will remain high. In 2017, the

Table 6.1 Top 10 cybersecurity businesses in different regions

Region	Number of businesses	Percentage (%)
Beijing	957	35.70
Guangdong	336	12.53
Shanghai	279	10.41
Jiangsu	143	5.33
Sichuan	133	4.96
Zhejiang	117	4.36
Shandong	110	4.10
Fujian	90	3.36
Hubei	72	2.69
Liaoning	69	2.57

Source China Academy of Information and Communications Technology

(Unit: 0.1 billion yuan)

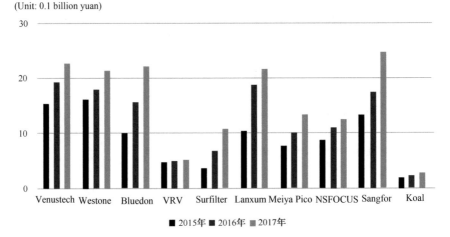

Fig. 6.3 Revenue of China's top 10 listed security businesses (2015–2017). *Source* China's domestic listed security businesses

average R&D expenditure of the 10 businesses was 237 million *yuan*, with a year-on-year increase of 23.43%; their average R&D investment growth rate was 35.37%.[8] Figure 6.4 shows the net profit and its growth rate of China's top 10 domestic listed security businesses from 2015 to 2017.

---

[8] *Source* China's domestic listed security businesses, *White Paper of China's Cyber Security Industry* (2018).

(Unit: 0.1 billion yuan)

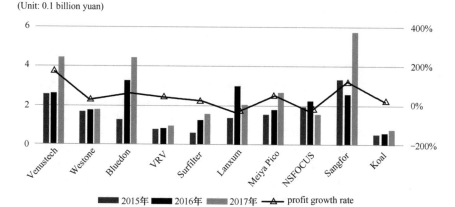

Fig. 6.4 Net profit and its growth rate of China's top 10 domestic listed security businesses (2015–2017). *Source* China's domestic listed security businesses

6.4.2 Development of Cybersecurity Technologies

1. AI facilitating cybersecurity technology innovation

With the multiplication of online attacks and the increase of their frequency, it is becoming increasingly difficult to protect cybersecurity. Automatic or intelligent solutions are sought after, with more advantages in AI technologies, some of which have emerged, including online behavior identification, active defense, timely warning, and automatic response and handling, which play an important role in applications. Notably, AI's application in cybersecurity protection is a two-edged sword, since it can be used to improve the efficiency or capability of online attacks. Therefore, AI technologies will see more innovative applications in cybersecurity protection.

2. New development of data security technologies

Data can be taken as "gold" or "oil" in cyberspace, becoming businesses' core assets and key innovative elements as well as strategic national resources. A large number of innovative applications based on data are bringing about great changes in such areas in China as health and medical care, finance and business, logistics and express, urban administration, social governance, and manufacturing. Data are the nuclear resources in cyberspace and the prior target of most online attacks. From 2017 to 2018, China's data security technologies have been developing at a high speed. Their focus has shifted from data security monitoring and disclosure prevention to protection of data sharing and non-structural databases as well as data disclosure tracing technology. Privacy protection technology is developing fast, aimed to solve the conflict between big data application and personal information protection.

3. **More security technologies developed based on cloud**

Innovative application of security technologies has been facilitated by the steadily fast growth of cloud security market volume and governmental regulation through policies. In July 2018, Forrester, an international investigation institution, issued *The Forrester Wave: Full-Stack Public Cloud Development Platforms in China*, which shows that the cloud platform is the foundation for decision-makers like CIOs and CTOs to embrace emerging technologies and accelerate value delivery. Big data, machine learning, and AI service based on cloud play a key role in the whole client life cycle. In the same month, Ministry of Industry and Information Technology issued *Implementation Guide for Businesses' Access to Cloud (2018–2020)*, which requires businesses to use security services like on-the-cloud host security protection, online attack protection, application firewall, key/credential management and data encryption. In 2018, CASB technology was promoted, and CWPP function was diversified. More cloud service suppliers have improved their ability to detect, their efficiency in response, and their ability to handle threats through threat intelligence sharing. Besides, more advanced authentication technology has been applied to cloud service to help the latter to stop more attackers from disguising themselves as legal users.

6.5   Cybersecurity Prevention, Protection, and Guarantee

6.5.1   Critical Information Infrastructure Security Protection

In April 2018, General Secretary Xi Jinping pointed out at the National Working Conference of Cybersecurity and Information Technology that accountability for critical information infrastructure security protection should be performed, relevant industries and businesses should be the main bodies responsible for such protection since they are the operators, and competent authorities should supervise the protection. In August 2018, to improve the protection of telecommunication and Internet industry network security, Ministry of Industry and Information Technology carried out a relevant inspection. Through recorded ranking, self-examination, and random examination, public networks can be operated safely and stably.

Since July 2018, provinces (and municipalities) such as Tianjin, Shanghai, Chongqing, and Shanxi have launched critical information infrastructure security inspection. They start from the critical business concerning national welfare and people's livelihood and identify the information system and industrial control system that may affect the operation of key business. By accurately assessing the security of China's critical information infrastructure, they can evaluate cybersecurity risks, so that they can facilitate administration, prevention, improvement, and construction through inspection and provide basic data for the construction of a guarantee system of critical information infrastructure security.

6.5.2 Classified Protection of Cybersecurity

According to Article 21 of *Cybersecurity Law of the People's Republic of China* (*Cybersecurity Law* for short hereinafter), China should implement classified protection of cybersecurity. Network operators should, in accordance with the law, perform their duties for security protection, and ensure that the networks will not be disturbed or damaged or visited without authorization, so that data will not be disclosed, stolen or falsified. On June 27, 2018, public opinions were collected for the *Rules of Classified Protection of Cyber Security (Draft for Opinions)*, which marked progress in the classified protection. In January 2018, National Information Security Standardization Technical Committee issued *Guide for Classified Protection of Cyber Security and Information Security Technology (Draft for Opinions)*, in which the principles, objects, and programs for classified protection are stipulated.

6.5.3 Security Protection of Industrial Internet

The Central Committee of CPC and the State Council attach importance to the security of industrial information and the construction of the industrial information security guarantee system, having launched a series of policies, founded supporting organizations for national industrial information security to develop the industrial information security industry, and issued a series of guiding regulations to produce talents in that field. In December 2017, Ministry of Industry and Information Technology issued *Action Plan for Industrial Control System Information Security(2018-2020)*. It is planned that by 2020, a full-system industrial security administration system will have been built and the capability of industrial innovation and development will have been improved dramatically. In May 2018, the first National Industrial Security Conference was held, at which the trend, standardization, practice, and talent production of industrial security were discussed. It was a world-level platform for exchanges and cooperation concerning industrial security technology, which has contributed to the development of China's business of industrial security.

6.6 Cybersecurity Talents Production and Promotion

It is the foundation of cybersecurity to increase cyber secure awareness, and produce talents in that respect. In recent years, cybersecurity awareness has been dramatically increased and a number of talents have been produced.

6.6.1   Accelerating Cybersecurity Talent Production

General Xi Jinping attaches importance to cybersecurity talent production, pointing out that the competition in cyberspace is the competition of talents. In recent years, China has sped up cybersecurity talent production. In February 2018, the General Office of Ministry of Education issued *Notice of Issuing Main Points for Education Information Technology and Cyber Security*, according to which cybersecurity talent production and cybersecurity protection will be improved, more degree authorization centers will be founded, and first-class cybersecurity institutes will be established. In June 2018, the Higher-learning Department of Ministry of Education issued *Notice of Undergraduate Major Establishment in Higher Learning Institutions*, requiring that majors like cybersecurity should be established in higher learning institutions to serve the country's strategic development. China Internet Development Foundation has set up a special fund for cybersecurity to accelerate talent production. In September 2018, 10 excellent talents and 10 excellent teachers in that respect were selected, with RMB 500,000 and RMB 200,000 for them, respectively, as the award. For 3 years after its establishment, one outstanding talent, 30 excellent ones, 28 excellent teachers, and 414 students have been awarded by the foundation.

6.6.2   Cybersecurity Promotion and Awareness Cultivation

1.   Continuous National Cybersecurity Publicity Weeks

On September 19, 2018, the Fifth National Cybersecurity Publicity Week was opened in Chengdu. It has been held every year since 2014. For the fifth of it, the theme was "cybersecurity for the people and by the people". A series of forums were held, including Cybersecurity Expo, Cybersecurity Technology Summit Forum, Critical Information Infrastructure Security, Big Data Security and Personal Information Protection, Internet Users' Quality Education, International Cybersecurity and Qualification Assessment, Integrated Development of New Economy and Cyber Security, Cyber Security Standards and Industry, New Cybersecurity Technology Development, and Cybersecurity Talent Production. Competent ministries organized thematic activities like Campus Day, Telecommunication Day, Law Day, Finance Day, Juveniles' Day, and Personal Information Protection Day. The fifth National Cybersecurity Publicity Week involved a number of participants with a wide coverage of information, contributing to the improvement of the people's awareness of cybersecurity and its prevention and protection.

2.   Study, publicity, and implementation of *Cybersecurity Law*

Since its promulgation, *Cybersecurity Law* has been publicized, interpreted, and studied through newspapers, journals, broadcast stations, TV stations, portal websites, and governmental WeChat/Weibo accounts. Its publicity has been enhanced in critical organizations and industries. For instance, Ministry of Industry and Information Technology has channeled the study of the Decisions and the Law[9] into the annual assessment indicators for basic telecommunication operators and organized key Internet businesses like Baidu, Alibaba, and Tencent to study the two documents together. Ministry of Public Security has organized staff from public security authorities throughout the country, central ministries and businesses, and over 260 cybersecurity businesses to study them together. The former State Administration of Press, Publication, Radio, Film, and Television launched cybersecurity knowledge and skill drilling and competition. Backbone staff from key organizations and industries in charge of cybersecurity in Inner Mongolia and Heilongjiang were trained. In Guangdong, Fujian, and Jiangxi, cadres attended workshops on cybersecurity and information technology to implement the *Cybersecurity Law* and *Provisions for the Administration of Internet News Information Services* and other laws and regulations concerning the Internet, so that cadres, staff in charge of key websites and professional backbones can know better about laws and hence implement them. Meanwhile, juvenile Internet users are the targets of law publicity. Activities like Cyber Security at School and in Family and To Be Good Internet Users have been launched, so that juveniles can access the Internet in accordance with laws.[10]

3. Diverse cybersecurity conferences and competitions

Since October 2017, there have been a variety of cybersecurity conferences and competitions, in which the entire society has been encouraged to participate to acquire knowledge and skills concerning cybersecurity. In that very month, the Second Mainland-Hong Kong Cybersecurity Forum was held in Xiamen; in January 2018, the Second Qiangwang Cup National Cybersecurity Challenge was launched; in March 2018, the Eleventh National College Students' Information Security Contest was held; in May 2018, the First National Industrial Information Security Conference was held in Beijing and meanwhile, the Industrial Information Security Skill Contest of 2018 was launched; in August 2018, China's Annual Cybersecurity Conference of 2018 was held in Beijing, with the theme of "Security Brain and Intelligent Ecosystem", and the Cybersecurity Contest of the same year was launched; and in September 2018, Internet Security Conference (ISC) of 2018 was held in Beijing, with the theme of "Security from Zero". All these cybersecurity conferences and competitions have been platforms for exchanges and discussions in terms of cybersecurity technologies, business, and administration and played a key role in producing talents and improving public awareness in that respect. To better hold cybersecurity competitions, Cyber Administration of China and Ministry of Public Security jointly issued in June 2018 the *Notice of Regulating Cybersecurity Competitions*, in which a series of regulations were provided.

---

[9]*Decisions of the Standing Committee of the People's National Congress on Enchancing Online Information Protection* was reviewed and adopted in December 2012 and the *Cybersecurity Law of the People's Republic of China* was reviewed and adopted in November 2016. The two documents are called the Decisions and the Law for short.

[10]*Source* http://npc.people.com.cn/n1/2017/1225/c14576-29726949.html, December 25, 2017.

Chapter 7
Improved Construction of Rule by Law for the Internet

7.1 Overview

Since the 18th National Congress of the CPC, the construction of China's cyberspace legislation has been improving. Under the guidance of the country's outlook on security, the Internet has been developing on the principle of cyber governance with the rule by law, following the trend of digital economy development and the combination of security and development. In general, policies have been adopted to promote development, and laws have been made to guarantee security.

(1) China, based on the *Cybersecurity Law*, is accelerating the legislation on the Internet, covering network information service, cybersecurity protection, and cyber administration, and is forming a preliminary legal framework concerning the Internet.
(2) The country has been advocating cyber administration, opening up and access in accordance with laws, and enhancing punishment over illegal information and behavior in cyberspace, which has made cyberspace governance better and cyberspace cleaner.
(3) The country is innovating the legal system for the Internet, and accelerating the founding of pilot Internet courts. Progress has been made in innovative case handling, court information technology development, and multiple dispute settlement mechanism building.

7.2 Steady Progress in Internet Legislation

China's Internet legislation is basically in line with the development of Internet development. In 1994, the country's computer information network was officially connected with the Internet. Since then, legislation in that respect has never stopped. The top-level design of the legal system concerning the Internet was completed after the 18th National Congress of the CPC. In the beginning, the legislation in that respect

© Springer Nature Singapore Pte Ltd. 2020
Chinese Academy of Cyberspace Studies, *China Internet Development Report 2018*,
https://doi.org/10.1007/978-981-15-4043-1_7

only covered single Internet administration. So far, it has been expanded into network information service, online platform management security protection, and social administration of cyberspace. Based on the *Constitution* and the *Cybersecurity Law*, legislation has been made in terms of information network infrastructure, network information service provision, online information, and cyber administration, which have all seen progress in legislation in the past year.

7.2.1 Network Information Service

Network information service is the essential content of China's Internet administration, widely applied in all areas with increasing significance. With the *Nine Prohibitions* as the basic standards, legislation on network information service provides regulations on online production and information spread. In the past year, the country has accelerated legislation on network information service, having formulated and launched regulations on news information service, cyberspace dedicated to heroes and martyrs, etiquette on Weibo, and programs for juveniles, to guide information service providers and Internet users in making cyberspace cleaner and cleaner.

1. **Improvement of news information service administration**

To enhance the administration over content management practitioners in Internet news information service providers and safeguard the legal rights and interests of the practitioners and the public, Cyber Administration of the People's Republic of China issued on September 30, 2017, the *Measures for the Administration of Content Management Practitioners in Internet News Information Service Providers*, which contains codes of conduct for practitioners in Internet news information service provision. On October 30, 2017, the same office launched *Provisions on the Administration of the Safety Assessment of New Technologies and Applications for Internet News Information Services*,to regulate the safety assessment of new technologies and applications for Internet news information service. The document provides that Internet news information service providers should establish and improve the safety assessment and guarantee system for new technologies and applications, carry out safety assessment independently in accordance with regulations, and improve in time the necessary information safety guarantee system and measures.

2. **Formation of cyberspace dedicated to heroes and martyrs**

On April 27, 2018, at the Second Session of the 13th Conference of the Standing Committee of the National People's Congress, *Law of the People's Republic of China on the Protection of Heroes and Martyrs* was passed to prohibit any twist of history or any spoof of heroes and martyrs. It requires that radio and TV stations, newspaper and journal publishers, and Internet information service providers spread stories of heroes and martyrs by launching works, advertisements, and columns about heroes and martyrs. It prohibits any twist, abuse, or negation of the deeds and spirit of heroes and martyrs, whose names, fame, and honors are protected by laws. All network operators are required to stop any spread of humiliation and slandering of heroes' and martyrs' names, reputations, and honors and eliminate the spread of information in time.

3. **Construction of micro-blog etiquette**

In recent years, some micro-blog service providers have seen their awareness of security accountability weakened, and their management measures and guarantee capacity unimproved. Thus, illegal and harmful information has spread, which causes damage to the legal rights and interests of citizens, legal persons, and other organizations and affects the order of information spread in cyberspace. To promote the healthy and orderly development of micro-blog information service, Cyber Administration of the People's Republic of China issued on February 2, 2018, the *Provisions on the Administration of Micro-blog Information Service*, which gives complete and specific provisions on platform qualifications, entity responsibility, identity verification, hierarchical and classified management, information security guarantee, rumor refuting mechanism establishment and improvement, industrial disciplining, and credit system establishment. It is an embodiment of the application of Internet legislation in the administration of micro-blog service.

4. **Reinforcement of programs for juveniles**

To bring the administration of programs for juveniles under the jurisdiction of laws, guide and regulate the content of the programs, and hence protect the legal rights and interests of juveniles, former National Radio and Television Administration drafted *Regulations on Programs for Juveniles (Exposure Draft)*, which requires that commercialized, and too adult and entertaining programs are prevented for juveniles, that programs for that group should not promote underage star effects or packaging, that no stars' children should be made to cause a sensation, and that no child under 10 should be an ambassador for any brand. No advertisement for medical treatment, medicine, health products, cosmetics, alcohol, surgery products, and online games should be broadcast through channels or websites for the underage and no such advertisement should be made before or after any program for juveniles.

7.2.2 *Cybersecurity Protection*

Since the 18th National Congress of the CPC, China has been attaching great importance to cybersecurity legislation. *Decision of the CPC Central Committee on Major Issues Pertaining to Comprehensively Promoting the Rule of Law* proposes that laws and regulations should be improved in terms of cybersecurity protection. Since the enforcement of *Cybersecurity Law*, the country has accelerated the legislation on cybersecurity, including the improvement of some supporting laws and regulations to meet the challenge to cybersecurity protection and to ensure better implementation of the *Cybersecurity Law*. In the past year, a series of regulations have been launched for personal information protection and classified protection of cybersecurity.

1. Personal information protection being enhanced

On December 29, 2017, Office of the Central Leading Group for Cyberspace Affairs, General Administration of Quality Supervision, Inspection and Quarantine of the People's Republic of China, and National Information Security Standardization Technical Committee (NISSTC) jointly launched *Information Technology—Personal Information Security Specification* (GB/T 35273-2017, *Security Specification* for short hereinafter). As a national standard, it is a supporting specification in terms of personal information security in the implementation of *Cybersecurity Law*, providing principles and requirements for security, and the information security administration system for businesses. On August 27, 2018, the draft of different parts of the *Civil Code of China* was submitted for review to the Fifth Session of the Standing Committee of the 13th National People's Congress. Against the problems in privacy and personal information protection, the draft reinforced the protection, laying a foundation for the personal information protection law to be formulated.

2. Classified cybersecurity protection

To facilitate classified cybersecurity protection and guarantee national cybersecurity and critical infrastructure security, Ministry of Public Security, Cyberspace Administration of China, State Secrets Bureau of the People's Republic of China, and State Code Administration jointly launched on June 27, 2018, *Classified Cyber Security Protection Rules (Exposure Draft)* as a supporting document of the *Cybersecurity Law* to handle the serious situation of cybersecurity and the problems in that respect. Contributing to the consolidation of the legal system of China's cybersecurity guarantee, it provides specific and operable requirements for the scope of classified cybersecurity protection, responsibilities of supervising authorities, obligations of network operators, and construction of classified cybersecurity protection. It is a legal document for classified protection.

7.2.3 Legislation in Economy and Society

In the past year, in line with the new application and new progress of the Internet and IT in economy and society, China has formulated laws and regulations on e-commerce, online car-hailing, and Internet finance, which can be seen as great progress in legislation.

1. E-commerce Law of the People's Republic of China

On August 31, 2018, the Fifth Session of the Standing Committee of the 13th National People's Congress reviewed and passed the *E-commerce Law of the People's Republic of China* (*E-commerce Law* for short hereinafter*)*, which is the country's first comprehensive law of e-commerce and digital economy, so it will be a key component in China's legal system of the digital economy. It contains 7 chapters and 89 articles, offering specific provisions on dealers, contract signing and implementation, dispute settlement, promotion, and legal responsibilities in terms of e-commerce. It is an open and proactive law, encouraging innovation and competition and covering regulation and administration. It is a framework for future e-commerce development of China.

2. **Online car-hailing supervision being enhanced**

To enhance the operation of supervision information interaction platforms for online car-hailing, regulate data transmission, improve online car-hailing supervision efficiency, and create favorable business environment, Ministry of Transport launched on February 26, 2018, *Regulations on Online Car-haling Supervision Information Interaction Platforms*, which requires that all transport authorities should enter the platforms of the information of online car-hailing businesses, cars, and drivers in time. From zero clock of the day after getting the license, car-hailing platforms should transmit to ministerial platforms the static information and operation data including information of order, business, location, and service quality. Subsequently, the ministerial platforms transmit the data in real time to provincial and municipal supervision platforms. The officially launched online car-hailing platforms will provide a data basis for the regulation of the car-hailing industry and facilitate the daily supervision of that industry.

3. **Financial activities on the Internet being regulated**

On December 1, 2017, Working Unit Office of Special Rectification of Internet Financial Risk and Working Unit Office of Special Rectification of P2P Online Lending Risks jointly issued *Notice of Regulating Cash Lending*, in which problems with cash lending are pointed out and a warning is given about potential financial risks and social risks. On July 11, 2018, People's Bank of China issued *Notice on Enhancing the Administration of Cross-border Financial Networks and Information Service*, which provides the obligations of overseas providers, including those in terms of pre-events, service, alteration, and emergency. Overseas providers are forbidden to construct exclusive financial networks to provide financial information transmission services within China. The notice also contains requirements for self-discipline within the industry.

7.3 Improved Internet Law Enforcement

General Secretary Xi Jinping pointed out at the Symposium on Cyber Security and IT Application that the cyberspace is no place outside of judicial reach and that we should enhance cyberspace governance and create clean cyberspace for Internet users, especially the juveniles, to show our responsibility for society and the people. A state cannot be ruled without law and the cyberspace cannot be developed without law. Cyberspace administration and governance is a law enforcement activity involving supervision. Without strict law enforcement, there would be no cyberspace governance in accordance with the law. Over the past year, cyberspace law enforcement authorities have enhanced the combat against illegal information and activities. High-profile cases have been dealt with, businesses that have breached the laws have been exposed, and hence the cyberspace law enforcement has been enhanced.

7.3.1   Clearance of Illegal Information on the Internet and Regulation of Online Communication Order

(1)   Centralized governance of online streaming and short video and spread of positive energy on the Internet

Cyberspace Administration of China, in collaboration with National Office against Pornography and Illegal Publications, Ministry of Industry and Information Technology, Ministry of Public Security, Ministry of Culture and Tourism, and National Radio and Television Administration, has launched a series of special governance actions against online streaming and short video, making every effort to get the low and obscene content out of the two online forms. Meanwhile, the production of excellent content of online streaming and short video is encouraged. In July 2018, the abovementioned six authorities carried out rectification of short video, urging all short video platforms to perform their accountabilities and strengthen self-examination and rectification. In August 2018, they urged local authorities to enhance online streaming licensing and recording as well as basic administration of that respect, establish the long-term mechanism, and carry out the clearing of illegal online streaming.

(2)   Clearance of illegal information spoofing and slandering heroes and martyrs and promotion of hero respect

Since *Law of the People's Republic of China on the Protection of Heroes and Martyrs* was enforced, the Supreme People's Court and Supreme People's Procuratorate have issued notices, requiring to punish, in line with laws, for illegal actions of abusing rights and interests of heroes and martyrs and spoof of their images. It is a long-term task to safeguard the dignity of heroes and martyrs and carry on their spirit. Only when we protect their rights and interests, can we facilitate the whole society's respect for heroes.

(3)   Clearance of information harmful for the underage to protect their healthy growth

To enhance the legal protection of the underage and make cyberspace clean, all competent authorities are reinforcing their supervision and have carried out Seedling-protection Initiative, combating all pornography and illegal publications concerning the underage. A number of high-profile cases have been tackled, which has contributed to the protection of juveniles' rights and interests.

7.3.2 Continuous Cybersecurity Examination and Implementation of Cybersecurity Law

1. **Enhancement of personal information protection and promotion of data industries development**

In February 2018, Ministry of Public Security called on all local public security authorities to carry out the Special Campaign: Cyber Cleaning 2018 to combat cybercrime chains in which some industries help with online fraud, pornography, and gambling. They tried every means to fight against upstream crime, especially, the provision of information, technology, and tool support for crimes of stealing personal information, hackers' attack, and illegal sales of "black cards". In September 2018, National Information Security Standardization Technical Committee (NISSTC) organized the Launching Conference of Special Working Businesses of Privacy Policy, at which privacy policies of five categories of businesses were reviewed and evaluated, including traveling and tourism, life service, film and TV entertainment, tool information, and online payment.

2. **Classified cybersecurity protection and its improvement**

Since the implementation of *Cybersecurity Law*, some cases of classified protection of cybersecurity have occurred in Shanxi, Guangdong, Shanghai, Sichuan, Chongqing, Anhui, Heilongjiang, and Hunan, involving fields like education, Internet, medical care, and public service. Punishment has been conducted to unclassified recording, and failure to make classified assessment, to keep online journals, to take safety measures, to adopt data classification, or to make backup and encryption for key data. In some cases, problems have been found by law enforcement authorities during their inspection while in most of them, the punished organizations have suffered vulnerability attacks and key information disclosure.

3. **Critical information infrastructure security protection and national cybersecurity safeguarding**

In 2018, Cyberspace Administration of China continued to carry out critical information infrastructure security inspection. Localities and competent authorities also launched the same inspection. They started from critical sectors closely related to the national economy and people's livelihood, cleared up information and industrial control systems that may affect critical business operation. They tried to learn about the condition of critical information infrastructure, and assessed cybersecurity risks to facilitate examination, prevention, improvement, and construction.

7.3.3   Sharpening of Law Enforcement and Promotion of Orderly Operation of the Internet

1. **Combat against online piracy and patent infringement and enhancement of intellectual property protection**

In July 2018, National Copyright Administration and Cyberspace Administration of China jointly launched the special action Sword Net 2018, which was aimed to combat online piracy and patent infringement. The campaign had online copyright infringement as the key goal, together with the investigation of cases. In it, classified supervision, interview and rectification, administrative punishment, and crime combat have been done, with the focus on the copyright infringement of online reproduction, short video and cartoons, and the administration concerning online streaming, knowledge sharing, and audiobooks platforms. Achievements have been consolidated in terms of online TV and movie broadcast and music, e-commerce, APP stores, and cloud memory to safeguard the order and copyright environment in cyberspace.

2. **Regulation on online game markets and accountability of businesses for society**

By the end of 2017, Public Relations Department and other seven ministries issued *Opinions on Regulating Online Game Markets Administration*. In line with the unified plan, all ministries collaborate with each other in inspecting and investigating high-profile cases concerning online game markets, promoting self-discipline within the industry, and hence creating clean cyberspace. Public security authorities of all levels combat crime concerning online games, having succeeded in detecting 98 crime cases. All provinces (and autonomous regions and municipalities directly under the Central Government) have launched the Initiative of Combating Pornography and Illegal Publications to deal with harmful political information in online games and pornography. In the campaign, obscene content has been removed from online games, and private servers (Internet servers obtaining installation programs without the authorization of copyright owners) and plugs (control of online game programs from outside) have been found out and removed.

3. **Supervision over online markets and safeguarding of market competition**

To carry out the decisions made in the 19th National Congress of CPC and *Plan of Market Supervision during the 13th Five-year Plan Period*, and to create an online market environment with honest business and fair competition, member organizations of Inter-Ministerial Joint Meeting System for Network Market Supervision decided to launch the Special Action of Online Market Supervision (New Sword Action) 2018 from May to November 2018. The Internet must be administered in accordance with laws through the Internet itself in credit and collaboration. Supervision has been optimized and its efficiency has been improved to combat online infringement and counterfeit, *shuadan* (false transaction) to make false reputation, false publicity and advertising, with the accountability for e-commerce platforms and contract format regulation as the focus for full-flow and full-chain supervision over online markets, so as to prevent crime in that regard, to improve online commodity and service quality and online market competition order and consumption environment, and to facilitate the construction of the social credit system.

7.4 Judiciary Innovation for the Internet

On June 26, 2017, the 36th Conference of Central Leading Group for Comprehensively Deepening Reforms passed the *Plan on the Establishment of Hangzhou Court of the Internet*. On August 18, 2017, Hangzhou Internet Court was put into operation. On July 6, 2018, the 3rd Conference of the Committee for Comprehensively Deepening Reforms reviewed and passed the *Plan on the Establishment of Beijing Internet Court and Guangzhou Internet Court*, to deepen the exploration in the establishment of Internet courts and to improve Internet case verdict. To regulate the lawsuit application to the Internet courts of Hangzhou, Beijing, and Guangzhou, and protect legal rights and interests of the parties concerned, the People's Supreme Court announced on September 7, 2018, the implementation of *Provisions of the Supreme People's Court on Several Issues concerning the Trial of Cases by Internet Courts*. All these measures are innovative attempts in the Internet administration of justice, and they are of significance to the governance of the Internet and to the concerned parties' involvement in lawsuit verdict in line with laws. Hangzhou Internet Court has been in the trial cooperation for one year, having witnessed progress in innovating case trials, speeding up court information construction, and establishing a pluralistic dispute settlement mechanism.

7.4.1 Innovation of Case Trials and Improvement of the Judicial Trial Mechanism

Hangzhou Internet Court has innovated its case trials, improved its judicial trial mechanism, reformed its case trial rules, and established a series of standards containing online lawsuit platform operation standards and online video trial regulations. They have succeeded in the control of trial procedures, standardization of time limit and procedures covering lawsuit application, trial and delivery, and improvement of trials by default. Moreover, they have made traceable the information on the litigation subject's identity. ID information of the parties involved in a legal case is first uploaded onto the Internet, then it is compared with the corresponding information on the ID database of public security authorities and it is rechecked during the video trial on the Internet. In this way, the authenticity of the ID information is guaranteed. More importantly, fast evidence gaining and cross-examination are guaranteed. The judge can introduce evidences on the Internet and check the facts involved in the case while the parties concerned can take online cross-examination at any time.

7.4.2 Reinforcement of Information Construction of Courts and Provision of Case Trial Convenience

The Internet courts of Hangzhou, Beijing, and Guangzhou have increased judicial efficiency and convenience through the application of IT and the establishment of the intelligent court system. Hangzhou Internet Court was the first court tackling Internet cases, embodying three features. First, through the online lawsuit tackling platform, the case can be tackled totally on the Internet, covering application, filing, trial, and execution, so that any Internet case can be handled on the Internet. The trial rate of that court is 100%, with a trial lasting only 25 min, and the interval being 48 days, up 50% in comparison with the traditional trial by a common court. Secondly, service is provided for litigant parties, who can just stay at home completing their procedure for litigation. Thirdly, the court, relying on the big data of judicature, can access the data of public security authorities, industrial and commercial administration, and e-commerce businesses point to point,so that it can share data with them, which is convenient for the parties involved in the case and judicial adjudication.

7.4.3 Regulation of Cyberspace Order and Construction of the Pluralistic Dispute Settlement Mechanism

Since it was put into operation, Hangzhou Internet Court has fulfilled the requirement for judicial reform and adjudication and combined Internet disputes with mechanism innovation. It has tackled disputes concerning online shopping and service and micro-finance lending as well as Internet intellectual property right infringement disputes and administrative cases concerning Internet management. The rate of verdict acceptance after the first trial reached 98.5%, which shows that the Court has contributed to safeguarding the cyber order and promoting justice and governance of the Internet by law. The pluralistic dispute settlement mechanism on the Internet has been facilitated. A dispute settlement system with Internet characteristics has been set up, involving multiple parties, good interaction, and coordinated mediation. Hangzhou Internet Court guides Internet businesses in pre-settling disputes and connects with some Internet businesses' platforms, taking them as the pre-settling programs and having them purify, discipline, and regulate themselves to release pressure on suing. Full-time mediators are employed for mediation. Committee for People's Mediation and Internet Society of China have their offices set up there, which has helped to promote the lawsuit withdrawal rate to 87% and the automatic implementation rate of mediated lawsuits to 100%.

7.4.4 Summarization of Lawsuit Judgement Rules for Improvement of Cyberspace Governance Capacity

The Supreme People's Court attaches importance to the summarization of Internet case tackling and verdict, dedicated to the exploration of verdict rules of new Internet cases, trying to solve any problem and safeguard cyberspace order through judicial approaches, promoting core socialist values, and contributing Chinese wisdom to the global Internet governance. In August 2018, it released the first typical cases concerning the Internet, including microfinance lending, online shopping, online service, privacy, intellectual property rights, and competition. The experience that the Supreme People's Court has summarized in terms of Internet cases in the past years provides some verdict clue for regulating online behaviors, preventing transaction disputes, and reducing transaction risks. The experience is also a reference for the people's court of all levels in dealing with Internet cases. It can be promoted among the people's courts of different levels, contributing wisdom and strength to the Internet case verdict. It is also a guide for the public in safeguarding their legal rights and interests and regulating Internet business practitioners' behaviors, thus guaranteeing the healthy development of the Internet economy.

Chapter 8
Active Participation in International Cyberspace Governance

8.1 Overview

International cyberspace governance is an important part of global governance. Adhering to the "four principles" and "five proposals", China keeps enriching the Chinese proposition of international governance of cyberspace. It gives full play to the roles of all entities, such as the government, international organizations, Internet businesses, technical communities, private institutions, and citizens, to realize common progress in cyberspace development, common maintenance of security, common participation in governance and sharing of results, and to work with other countries to build a community of shared future in cyberspace.

The Chinese government, social organizations, the private sector, technology communities, experts, and scholars have all participated in global Internet development and governance, formulation and R&D of global technical standards, and management of critical infrastructure resources of the Internet. The world Internet conference will be continued to contribute Chinese wisdom to global Internet governance.

China promotes diversified and multilevel international cooperation in cyberspace and strengthens cooperation with major countries, and collaboration in "digital Silk Road" construction with countries along the "Belt and Road", using the regional multilateral platform to strengthen interregional network cooperation and exchanges.

© Springer Nature Singapore Pte Ltd. 2020
Chinese Academy of Cyberspace Studies, *China Internet Development Report 2018*,
https://doi.org/10.1007/978-981-15-4043-1_8

8.2 China's Proposals on International Cyberspace Governance

8.2.1 China's Position in International Governance in Cyberspace as an Important Manifestation of the Concept of Extensive Consultation, Joint Construction, and Shared Benefits in Global Governance

The world is further multi-polarized, economically globalized, socially informatized, and culturally diversified. The reform of the global governance system and international order is being accelerated. Countries are becoming more interconnected and interdependent. The international power is more balanced, and the peaceful development trend is irreversible. Chinese President Xi Jinping has systematically elaborated on China's new ideas and thoughts on global governance and has proposed new solutions to essential issues of global governance in order to promote the development of the global governance system toward a more just and rational direction. The country upholds the principle of extensive consultation, joint construction, and shared benefits in global governance. It also advocates democracy in international relations, and believes that all countries, big and small, strong and weak, rich and poor, are equal. China supports the United Nations in playing an active role and supports the expansion of the representation of developing countries in international affairs.

Cyberspace governance is an important part of global governance. China practices the global governance concept of jointly building and sharing, and participates in cyberspace governance. Faced with outstanding problems such as uneven development of global cyberspace, imperfect rules, and unreasonable order, Chinese President Xi Jinping put forward "four principles" for promoting the reform of the global Internet governance system and "five proposals" for building a community of shared future in cyberspace. For the first time, the "Chinese position" on Internet Governance was fully and systematically expounded. On December 3, 2017, President Xi sent a letter to congratulate on the opening of the Fourth World Internet Conference. He emphasized that China advocates the "four principles" and "five proposals". That is, we hope to work with the international community to respect the sovereignty of the Internet and promote the spirit of partnership. According to him, everything will be discussed by all countries, so that we can promote development, maintain security, participate in governance, and share the results together. China will continue to play the role of a responsible big country, participate in the reform and construction of the global governance system of cyberspace, and continuously contribute its wisdom and strength.

8.2.2 Respect for Cyber Sovereignty as the Solid Foundation for International Cooperation

Respect for the sovereignty of countries in cyberspace is one of the core concepts of China's efforts to reform the global Internet governance system and build a community of shared future in cyberspace. In the congratulatory speech to the First World Internet Conference, President Xi Jinping made it clear that "cyber sovereignty should be respected". In 2015, he attended the Second World Internet Conference and delivered a speech there, taking "respect for cyber sovereignty" as the most essential one of the "four principles" for promoting the reform of the global Internet governance system. He stressed that the reform of the global Internet governance system could be promoted and a peaceful, secure, open, cooperative, and orderly cyberspace could be jointly built only on the basis of respect for each other's cyber sovereignty, on the principle of noninterference in other countries' internal affairs, and equality and mutual benefit, and for the goal of building a community of shared future in cyberspace.

In the information age, cyberspace has become the fifth frontier beyond land, sea, air, and sky. As President Xi Jinping has pointed out, the basic norms of international relations with the UN Charter as its core also apply to national sovereignty over cyberspace. This proposition has been widely recognized by the international community. Respect for the cyber sovereignty of all countries has become a solid foundation for international cooperation in cyberspace. In the international governance of cyberspace, the right of countries to choose their own development path, network management model, Internet public policies, and equal participation in international cyberspace governance should be respected. Drawing on international experience, all countries have the right to formulate Internet policies applicable to their national conditions. Meanwhile, they should also fulfill their corresponding obligations. They are not expected to engage in, connive at, or support network activities that endanger other countries. No country is allowed to use the network to interfere in the internal affairs of other countries or harm the interests of other countries.

8.2.3 Building of a Community of Shared Future in Cyberspace: An Irresistible Trend

It is a critical period for the reform of the global Internet governance system. Building a community of shared future in cyberspace has increasingly become a consensus of the international community. The rapid development of the Internet and information technology has not only injected strong impetus into economic and social development, but also brought many new challenges to sovereignty, security, and development of all countries. At a critical stage of the global Internet governance system change, no country can deal alone with the challenges faced by human beings and no country can retreat into an isolated island. Cyberspace is the common home of human beings, so it should be controlled by all countries, who should strengthen communication, expand consensus, deepen cooperation, and build a community of shared future in cyberspace. Only when countries develop together can the development dividend of the digital economy be fully released; only by working together can we cope with the increasingly serious security threats and challenges such as cybercrime and cyberterrorism.

The content of a community of shared future in cyberspace has been enriched in practice. First and foremost, a community of shared future is a security community. Countries should work together to address common security threats such as cybercrime and terrorism. More importantly, they should work together to build peaceful cyberspace, oppose arms race in cyberspace, and resolve conflicts and disputes by peaceful means. A community of shared future is also a community of development. Economic and social development is the vision of all countries. We should promote cooperation in the digital economy, facilitate the flow and sharing of capital, technology, information, talents and other factors related to cyberspace, and work together to strengthen energy development, and achieve mutual benefit and development. A community of shared future is a cultural community. We should foster positive common values and promote exchanges and mutual learning among the fine cultures of all countries. The diversity of civilizations should not be a source of conflict. Harmonious, diversified, and inclusive cultural exchange is beneficial to the world's peace and development.

8.2.4 Open Cooperation to Promote Transformation of the Global Internet Governance System

We should adhere to multilateral and multiparty participation in the international cyberspace governance, and allow the government, international organizations, Internet businesses, technical communities, private institutions, and individual citizens to play their main roles. It is necessary to promote network governance within the UN framework and to give full play to the role of various non-state actors.

Interconnection is the basic attribute of the Internet and the fundamental guarantee for maintaining the vitality and continuous development of the Internet. Strengthening open cooperation in cyberspace is an important goal in building a community of shared future in cyberspace. It is necessary to promote bilateral, regional, and international cooperation at different levels on an open basis, jointly build a platform for communication and cooperation, strengthen resources complementation, and achieve common development.

8.3 Diversified and Multilevel International Cooperation in Cyberspace

8.3.1 Bilateral Exchanges and Dialogues

1. China and the United States

China and the United States have both cooperation and differences on issues such as the Internet governance model, digital economic rules, and codes of conduct. In October 2017, the first U.S.–China Law Enforcement and Cybersecurity Dialogue was held in Washington, D.C. The two sides reviewed and summarized the achievements of their cooperation in law enforcement and cybersecurity in recent years. They also had in-depth exchanges on terrorism combat, drug control, cybercrime countering, fugitives chasing and stolen assets recovery, and illegal immigrants repatriation. They agreed to follow the important consensus reached at the Mar-a-Lago meeting between President Xi Jinping and President Trump, according to which they would adhere to mutual respect, legal reciprocity, honesty, and pragmatism, give full play to the role of U.S.–China Law Enforcement and Cybersecurity Dialogue, and strengthen bilateral dialogue and cooperation on law enforcement and cybersecurity. In 2018, the U.S. government issued the *National Security Strategy, National Cyberspace Security Strategy, National Cyber Strategy,* and other documents, identifying China as a rival, which may affect the cooperation and exchanges between the two countries in cyberspace.

2. China and the EU

In May 2018, the fourth meeting of the China–EU Digital Economy and Cybersecurity Expert Working Group co-sponsored by Cyberspace Administration of China and the European Commission's Directorate General for Communication Networks, Content and Technology was held in Antwerp, Belgium. Experts from both sides had an in-depth discussion on the digital economy science improvement/data use and protection/consumer protection in the digital age/digital economy innovation environment, artificial intelligence/blockchain and digital finance challenges, 5G applications, smart cities, and other fields before reaching a consensus.

In September 2018, the Ninth China–EU Information Technology, Telecommunications, and Informatization Dialogue was held in Beijing. The two sides reviewed the progress of their cooperation in the field of information and communication since the eighth dialogue meeting, focusing on issues such as information and communication technology (ICT) policy and digital economy, ICT supervision, 5G research and development, and industrial digitalization. The two sides state that they have broad common interests and great potential for cooperation in the field of information and communication. They will make full use of China–EU information technology, telecommunications, and information dialogue mechanism to strengthen policy communication and mutual understanding, promote credit enhancement, and expand cooperation in the field of 5G applications, industrial Internet, and AI.

3. China and Russia

The China–Russia comprehensive strategic partnership of cooperation has become deeper and more pragmatic. Cooperation and exchanges between the two countries in cyberspace have also been promoted, and they are increasingly moving from strategic to pragmatic cooperation. On June 8, 2018, they signed the *Joint Statement of the People's Republic of China and the Russian Federation*, emphasizing the expansion of exchanges between the two countries in information and communication technology, and the digital economy. The statement also stresses the improvement of information and communication infrastructure interconnection, the reinforcement of cooperation in radio frequency and satellite orbit resource management, promotion of the development of information network space, and deepening of mutual trust in the field of cybersecurity. In September 2018, during President Xi Jinping's visit to Russia to attend the 4th Eastern Economic Forum (EEF), presidents from two countries attended the China–Russia Local Leaders' Dialogue to promote the development of cross-border e-commerce between the two countries.

4. China and Germany

Significant progress has been made in cooperation and exchanges between China and Germany. In May 2018, the cybersecurity consultation under the China–Germany high-level security dialogue was held in Beijing. The two sides exchanged views on the cybercrime situation, cybercrime and security legislation, and cybercrime and cyberterrorism combating. They agreed to jointly promote cybersecurity law enforcement cooperation within the framework of the Sino–German high-level security dialogue mechanism. In July 2018, the fifth round of Sino-German government consultation was held in Berlin. According to the joint statement, "The two sides agree that bilateral cybersecurity consultation is the core platform for discussing cybercrime and cyber security cooperation. At the same time, we can use this consultation mechanism for exchange in economic influences, especially the risks and challenges posed by the network to data security and intellectual property protection, trade infringement and trade secrets. Given the important role that data storage, use and protection play in the core of future industry, both countries will provide protection for confidential data protection and secure cross-border data transmission when developing and implementing our own network security regulations."

8.3.2 Serving the Belt and Road Initiative

Chinese President Xi Jinping stresses that it is necessary to take the "Belt and Road" construction as an opportunity to strengthen cooperation with countries along the route, especially developing countries, in network infrastructure construction, digital economy, and network security to build a digital Silk Road in the twenty-first century. At present, China and 12 countries along the "Belt and Road" have achieved direct connectivity through cross-border land cables and international submarine cables. The country also intends to cooperate with international organizations such as the International Telecommunication Union to promote multilateral cooperation initiatives such as the East African Information Highway and the Asia-Pacific Information Superhighway. China's information and communication businesses have participated in the construction of information and communication infrastructure in more than 170 countries around the world, laying the foundation for the construction of the "Belt and Road" and the online Silk Road.

In September 2017, the 2017 Online Silk Road Conference was successfully held to promote new progress in pragmatic cooperation between China and Arabian countries. In the past 2 years, China and Egypt, Saudi Arabia, and other Arab countries have jointly promoted the construction of the online Silk Road, and carried out pragmatic cooperation in the fields of information infrastructure, cross-border e-commerce, smart cities, etc. The country has seen positive results in information interconnection. For instance, it has steadily promoted the "Silk Cable Project" within itself and countries like Kyrgyzstan, Tajikistan and Afghanistan, and "Fazabad-Wakhan Corridor-Kashi Cable Network Project" in Central Asia. It is promoting cooperation with Arabian countries on BeiDou Navigation Satellite System and developing cooperation and exchange in information infrastructure, satellite application services, big data, cloud computing, smart city, and other emerging areas. In addition, cross-border e-commerce has become a new engine for cooperation between the two sides. Businesses from China and Arabian countries are collaborating with each other in trade information technology service platforms, cross-border payment Internet platforms, and overseas logistics and warehousing of China's famous, high-quality and special products.

In September 2017, representatives of the Chinese and Bangladeshi governments signed a framework agreement in Dhaka, the capital of Bangladesh. China will provide preferential loans to Bangladesh to support the construction of the Bangladesh government's infrastructure network phase III project and communication network modernization project. The two projects are the infrastructure carriers of the "Digital Bangladesh" strategy and will directly drive the development of communications, information technology, government, and education in Bangladesh and promote social and economic development. At the same time, they will help improve interconnection and communication between China and Bangladesh in information and communication infrastructure. They are an important part of the "Belt and Road" construction.

In June 2018, China and Kazakhstan signed a joint statement, which made it clear that the two countries would jointly fight against cybercrime. In the same month, presidents from both countries signed the MOU on e-commerce cooperation, stressing that they will establish an e-commerce cooperation mechanism, promote e-commerce cooperation along the Silk Road, strengthen experience sharing, conduct personnel training, promote government–business dialogues and support businesses of the two countries in e-commerce cooperation.

In September 2018, the third China–ASEAN Information Port Forum was held in Nanning of Guangxi Province. With the theme of "Building a Digital Silk Road and Sharing the Digital Economy", the forum focused on the building of a digital Silk Road, sharing of new opportunities for digital economy development and new achievements in information technology, suggestion on creating a happy new life and cooperation based on the consensus of win-win cooperation to promote the implementation of the "Belt and Road Initiative" and to accelerate the construction of the China–ASEAN Information Port and the "Southward Passage".

8.3.3 Regional Multilateral Cooperation

As network issues are expanding into various fields, the existing regional multilateral cooperation mechanism has also begun to include network issues in the discussion and on the cooperation agenda. The Chinese government always attaches importance to and supports multilateral governance at the regional level, promotes the transformation of regional cooperation into digital cooperation, and incorporates cybersecurity and digital economy development into regional cooperation.

China attaches great importance to various cooperations under the framework of the BRICS countries and incorporates joint cybercrime combating into cooperation and accelerates the cooperation and exchange of cyberspace among BRICS countries. In September 2018, the BRICS leaders met for the tenth time and adopted the *Johannesburg Declaration*, which stated that the development of information and communication technologies have brought about indisputable benefits and new development opportunities, especially in the context of the fourth Industrial Revolution. However, the development gives rise to new challenges and threats, including the use of information and communication technologies for criminal activities whose number continues to grow, and increasingly serious abuse of ICTs by States and non-state actors.

The BRICS countries will strengthen international cooperation to combat terrorism and criminal activities through ICT and reaffirm the need to develop universally accepted legal instruments to combat ICT crimes within the framework of the UN. At the security level, the BRICS Roadmap of Practical Cooperation on Ensuring Security in the Use of ICTs or other consensus mechanisms will continue to be promoted, and it is proposed that relevant intergovernmental cooperation agreements should be made as to the BRICS cybersecurity cooperation framework.

The network cooperation between the member states of the Shanghai Cooperation Organization (SCO) has seen progress. In June 2018, leaders of the member states held a meeting of Council of Heads of SCO Member States and adopted the *Qingdao Declaration*, which states that "The dynamically developing world is currently going through a period of major changes and reconfiguration; the geopolitical landscape is becoming diversified and multipolar, and ties between countries are becoming closer. The Member States call on the international community to put more effort into creating a peaceful, secure, open and structured information space based on cooperation. They emphasize the central role of the UN in developing universal international rules and principles as well as norms for countries' responsible behavior in the information space and advocate the establishment of a working mechanism within the framework of the UN based on a just geographical distribution in order to develop standards, rules or principles for countries' responsible behavior in the information space and to formalize them by adopting the corresponding UN General Assembly resolution." "The Member States are convinced that all states should participate equally in internet development and governance. A governing organization established to manage key internet resources must be international, more representative and democratic". The Member States will continue to promote practical cooperation in countering threats and challenges in the information space based on the *Agreement Between the Governments of the SCO Member States on Cooperation in Ensuring Information Security* (Yekaterinburg, 16 June 2009), such as international cooperation in combating IT abuse, including the abuse for terrorist and criminal purposes, and call for developing an international legal document on countering IT use for criminal purposes under the auspices of the UN.

The Lancang-Mekong regional cooperation is a regional cooperation mechanism established by China, Thailand, Cambodia, Laos, Myanmar, and Vietnam for the purpose of strengthening cooperation among the countries along the Lancang-Mekong River. In January 2018, the first Lancang-Mekong cooperation leaders' meeting was held in Phnom Penh, Cambodia. They formulated a 2018–2022 action plan and put forward a number of practical cooperation agendas concerning cyberspace, including joint efforts to combat cyberterrorism, cybercrime, and other nontraditional security issues. They will also promote the construction and upgrading of information networks and other infrastructure, and increase the application of global satellite navigation systems, including the BeiDou Navigation Satellite System, in infrastructure construction, transportation, logistics, tourism, agriculture, and other fields in the countries along the Lancang-Mekong River. They formulated broadband development strategies and plans for countries along the river, committed to facilitating the construction and expansion of cross-border land and marine cables. They will also strengthen cooperation in innovation and development of digital TV, smartphones, smart hardware, and other related products.

8.3.4 Cooperation Encouraged Among Think Tanks, Businesses, and Other Nongovernmental Partnership

In the context of deepening bilateral and multilateral networks' exchange and cooperation, various types of think tanks, research institutions, and businesses play an indispensable role. China supports and encourages various nongovernmental organizations such as universities, research institutions, and corporate think tanks in participating in international exchanges and cooperation in cyberspace on various occasions. It is also enhancing mutual understanding and collaboration among think tanks, businesses, and other private organizations related to network governance in various countries around the world. On July 5, 2018, the Cyberspace Administration of China and Ministry of Education jointly organized the first "International Cyberspace Governance Research Bases". Ten universities, namely, Tsinghua University, Beijing University of Posts and Telecommunications, Wuhan University, Fudan University, Harbin Institute of Technology, Beihang University, Zhejiang University, People's Public Security University of China, Southeast University, and Tongji University, became the first bases in contributing to international cyberspace exchanges and cooperation and the development of cyberspace administration.

China Academy of Cyberspace, China Institute of Contemporary International Relations, and China Academy of International Studies cooperate with foreign think tanks and universities in holding bilateral and multilateral security dialogues. In December 2017, China Academy of Cyberspace and the Brookings Institution of the United States jointly hosted the World Internet Conference High-End Think Tank Forum, whose theme was "New Type of Great-power Relationship in Cyberspace". In April 2018, China Institute of Contemporary International Relations held a track II dialogue on China–EU cybersecurity. The dialogue has deepened the communication between China and other countries in cyberspace, and has created complementation for intergovernmental cooperation.

Looking into the future, China will seize the opportunities brought about by informatization, continue to promote the construction of network information infrastructure, accelerate the research and development of information technology, and fully release the vitality of digital economy development in a more open and inclusive way and through more practical and innovative measures. It will strengthen the construction of network content, improve cybersecurity, participate in international cyberspace governance, and promote the sustainable and healthy development of the Internet.

Afterword

The 19th National Congress of the Communist Party of China put forward the Two-Step Development Strategy, ushering in the new era of socialism with Chinese characteristics. Especially, the Congress made an important strategic layout for promoting the development of the Internet, big data, AI and sharing economy and for making China into a strong country in cyberspace, a digital country and an intelligent society, providing fundamental principles and guidelines for the development of the Internet. In the new era, we have the new march and new mission. It is our hope that *China Internet Development Report* Thoughts on Socialism with Chinese Characteristics, especially his strategic thoughts of strengthening and developing the country through the Internet, an exhibition of China's achievements in Internet development, a systematic summary of Chinse experience in Internet development and governance, and a scientific outlook on China's Internet development prospect, so that it can better boost the Internet development. We also hope that the *Report*, through a detailed study of China Internet, will contribute Chinese experience and wisdom to the world Internet development and governance.

The compilation of the *Report* has won support and guidance from Cyberspace Administration of China, especially guidance from the leaders of the Administration and support in terms of data and material from the departments and institutions of the Administration. China Academy of Cyberspace, as the coordinator, has invited a number of think tank institutions to participate in the compilation, including China Computer Network and Information Security Center, China Academy of Information and Communication Technology (CAICT), China Industrial Control Systems Cyber Emergency Response Team (CICS-CERT), State Information Centre, Peking University, Beijing University of Posts and Telecommunications and Xidian University. The participants in the compilation of the book are Yang Shuzhen, Fang Xinxin, Hou Yunhao, Li Yuxiao, Li Changxi, Liu Shaowen, Feng Mingliang, Xu Yunhong, Liu Bing, Jiang Wei, Chao Baodong, Li Zhigao, Tian Yougui, Long Ningli, Han Yunjie, Liu Yan, Dong Zhongbo, Wang Xiaoshuai, Ma Teng, Mu Chunbo, Yu Fengxia, Tian Li, Hu Jiarui, Sun Luman, Xie Yi, Yang Shuhang, Xiao Zheng, Li Wei, Chen Jing, Yuan Xin, Zhao Gaohua, Zhang Qiyuan, Wu Wei, Li Yangchun, Li Xiaojiao, Deng Yushuang, Shen Yu, Meng Qingshun, Long Chaoze, Wang Hualei, Yang Xuecheng,

© Springer Nature Singapore Pte Ltd. 2020
Chinese Academy of Cyberspace Studies, *China Internet Development Report 2018*,
https://doi.org/10.1007/978-981-15-4043-1

Liu Huailiang, Xu Yuan, Wang Xiaoqun, Fang Yu, Lang Ping, Wu Shenkuo, and Shen Yi.

The publication of the *Report* is also owed to the support and help from society. Due to our limited experience, capability and time for the compilation, there are probably errors in it. We sincerely hope that governmental agencies, international organizations, research institutions, Internet businesses, social associations, and people from all walks of life will offer their opinions and suggestions and provide more detailed materials so that we can make modifications and improvement to better support China Internet development.

China Academy of Cyberspace
October 2018